# How Computers Create Social Structures

Place, Community and Social structures

Silvio Carta

# How Computers Create Social Structures

## Accidental Collectives

**palgrave**
**macmillan**

Silvio Carta
School of Design
University of Greenwich
London, UK

ISBN 978-3-031-62851-1      ISBN 978-3-031-62852-8   (eBook)
https://doi.org/10.1007/978-3-031-62852-8

Cover illustration: Pattern © Harvey Loake

This Palgrave Macmillan imprint is published by the registered company Springer Nature Switzerland AG.
The registered company address is: Gewerbestrasse 11, 6330 Cham, Switzerland

If disposing of this product, please recycle the paper.

*To my parents, Alice and Stefano, thank you for everything.*

# ACKNOWLEDGMENTS

This book is the result of the last few years of work and studies on automation and data-driven approaches to spatial formation. During this time, I had the pleasure of working with many talented colleagues from whom I learned significantly and with whom I had many interesting discussions.

For the several stimulating projects on automated design, care homes and self-organising social spaces I would like to thank Ian Wyn Owen, Tommaso Turchi, Rebecca Onafuye, Luigi Pintacuda, Foteini Papadopoulou, Stephanie St Loe, Joel Simon, Davide Pisu, Miguel Vidal Calvet, Lubo Jankovic, and Luciano Ambrosini among many more extraordinary colleagues, students and friends. I would like to thank Farshid Amirabdollahian, Hongwei Wu, Catherine Menon and Christoph Salge for the stimulating discussions on AI and automation and Neil Spencer for the help with the statistical work.

The idea for this book originated during the 16th and 17th Annual International Conference of the Architectural Humanities Research Association in Dundee (2019) and Nottingham (2020), respectively. As such, I am in debt to all my great colleagues who helped me shape these ideas over fruitful conversations and debates. These include Professor Jonathan Hale and Professor Natasha Lushetich, Hazel Cowie, Alida Bata, Mary Ann Ng Yihui, Simone Chung, and many more.

This book builds on great work developed by many scholars in the past few years. In particular, I would like to mention Keith McNulty and his great dissemination work on data science, Grégoire Martinon and his work on algorithmic bias, and Luciano Floridi for his studies on digital ethics.

I would also like to thank Pieter Francois and Matthew Zook for their insightful feedback on the initial stages of this project.

A big thank you goes to all the staff at Palgrave Macmillan who believed in this project and had the patience to wait for my several postponements and delays.

Lastly, I would like to thank my family for their patience during long weekends and holidays over the past two years, when I often disappeared to work on this project.

London, September 2023.

# CONTENTS

# About the Author

**Silvio Carta** is an architect (ARB/RIBA), Chartered Building Engineer (MCABE) and Professor of Architecture at the University of Greenwich, UK.

His research includes Artificial Intelligence and Machine Learning design methods applied to the built environment, urban data science, data-driven approaches and computational design. He is currently the Section Editor of Computational Sustainability and Design, *City and Built Environment* (Springer/Nature), and a member of the technical committee: Data Sensing & Analysis (DSA) of the European Council on Computing in Construction (EC3).

Professor Carta is the author of *Big Data, Code and the Discrete City. Shaping Public Realms* (Routledge 2019), *Machine Learning and the City: Applications in Architecture and Urban Design* (Wiley 2022), and *The Physical and The Digital City: Invisible Forces, Data and Manifestations* (Intellect 2024).

# LIST OF FIGURES

CHAPTER 1

# Introduction

**Abstract** The idea of accidental collectives: groups of social media users created as a result of automated computer processes, is the core idea of this book. Accidentally Collectives is underpinned by the thesis that software unintentionally groups people through data analysis and classification as technology uses more complex sophisticated and powerful algorithms to automate routine tasks within social platforms. The book explores the stochastic character of these clusters, looking at how algorithms with predetermined objectives might build social networks around interests that may not necessarily reflect those of the users of such platforms. These accidental collectives are driven by forces within digital platforms that are invisible to users who find themselves into new and often unexpected digital spaces. The book examines the agency of software in social interactions and the degree to which the intentions of users and developers are represented in these automated processes through the analysis of case studies and real-life examples. Ultimately, it sheds light on the complex mechanisms behind the formation of new social groups in the digital age.

**Keywords** Accidental • Collectives • Automation • Data • Bias • Complexity • Transparency

© The Author(s), under exclusive license to Springer Nature Switzerland AG 2024
S. Carta, *How Computers Create Social Structures*,
https://doi.org/10.1007/978-3-031-62852-8_1

1

## Accidental Collectives

This book introduces the idea of accidental collectives: groups of social media users generated as a by-product of the automated work of computers. Software has a growing influence in our lives automating and optimising mundane, time-consuming and repetitive tasks. In doing so, groups of people are automatically created as the result of classification and data analysis. Once grouped by the invisible agency of software, people interact and establish new relationships, generating new collectives and communities. What is perhaps most interesting in the occurrence of this phenomenon is the relationship between the goals that underpin the algorithms and the actual group of people created. Most of the time, this relationship is stochastic (random, unrelated), yet in other cases there may be some pattern that may emerge, mostly driven by a form of bias. Let's imagine that a certain cluster algorithm (more on this in Chap. 2) is calibrated to maximise the connection between users with similar interests in a certain group of products. The algorithm will follow that principle prioritising choices to achieve its main objective. In this process, many possible options for connections among users will be neglected in favour of the most effective alternative that represent the best links between users with a share interest in that particular product. What is produced as the final outcome is a new network of people connected on the basis of interests they may have not even thought of having as a priority. In this new group, users find themselves associated with each other or suggested. Such groups may probably never happened if it wasn't for the algorithm that was clustering them as users with similar interest. In this lays the accidentality of these collectives, as groups formed by an invisible driving force running in the backend of the social media platform. Perhaps the main difference between the generation of such collectives within digital spaces and physical environments is the randomness. In the case of digital platforms associations with other users may appear as entirely random and perhaps even been perceived as such by individual users, or based on a clear factor (e.g., worked in the same company, educated in the same secondary school etc.). Such encounters may be perceived as coincidental as a those that we are familiar with when meeting an old school mate while walking on the street of a different town. However, we will see how such randomness is in reality the result of a precise objective function that drives the recommendation system of the digital platform.

Tangentially, we also look at the drivers that make such complex and well-developed mechanisms to work. I refer to these drivers in some part as forces, suggesting that the final result of the data (or social platform users) grouping is in a state of balance. Some of these forces are stronger than other, for example, in certain social media platforms, we can see certain patterns of clustering that favour users with similar interests, rather than others. Such *invisible forces* are subtle yet extremely powerful within the dynamics of what we see in social media every day.

To summarise, collectives are created as the precise and weighted outcome of a sets of algorithms that implement often competing objectives. Such algorithms are programmed to analyse available data and take decisions in autonomy. It is in this autonomous agency that we give to intelligent systems that new and interesting aspects emerge, when we look at the collectives are created (I expand on this idea in Chap. 3).

With the support of case studies and real-life examples, this book explores the accidental nature of this generation of new social groups and questions the role of software in the social interactions concerned. Accidental Collective also investigates the agency of software in social interactions, questioning if software has an intent and the extent to which the intentions of their developers, as well as of their users are directly or indirectly represented.

## THE RESEARCH FRAMEWORK FOR THIS BOOK

There is a growing body of work that analyse and discuss the role of computers in the generation of social groups. The idea of **digital self** has been studied by many. Notably we mention Danah Boyd's extensive work on youth and the use of social media (Ellison and Boyd 2013; Boyd 2014), where the author explored the identity that teenagers create within social media, their search for new public and groups, as well as other related aspects, including privacy, risk, bullying, inequality and possible addition to sites. Sherry Turkle's book in 2005 (Turkle 2005) provides us with a clear context for studies on the social implications of automation with computers. The author explored the mechanisation of the mind and the appearance of some new behaviours where users could think as machines as the unplanned consequence of a growing use of computers for games and social activities in young kids. More recently, some of these ideas have been expanded by Brett Frischmann, Evan Selinger in their Re-Engineering Humanity (Frischmann and Selinger 2018), with a lucid analysis of the ways in which individuals are changing behaviours in smart techno-social environments. References that provide a wider perspective on digital life include (Lupton 2015).

The **ethical side of automation** and algorithm-driven processes is covered in many recent studies. We note the Ethical Algorithm (Kearns and Roth 2019), Compromised Data (Elmer et al. 2015), Algorithmic transparency in media (Diakopoulos and Koliska 2017), Social Media Democracy (Persily and Tucker 2020), Automating Inequality (Eubanks 2018), Algorithms of Oppression (Noble 2018) and Weapons of Math Destruction: How Big Data Increases Inequality and Threatens Democracy (O'neil 2017) among many others.

The part of this study on **data** and their growing influence in social contexts has been heavily influenced by seminal texts like Data Revolution (Kitchin 2014), Data Lives (Kitchin 2021), materiality of information (Dourish 2022), data in society (Evans and Ruane 2019) and the idea of data derivatives (Amoore 2011).

Underpinning all these important studies, there is a plethora of more technical work that allows social media to exist and their continuous development. The majority of the decisions that are taken everyday in social media, both consciously and automatically, are governed by the logics of **network analysis**. This is an incredibly vast domain of expertise and, within the context of this book, we should simply acknowledge some key reference that readers can use to gather some initial insights and helpful information. Among the many texts, studies and handbooks on network analysis, we should mention the work of Keith McNulty, especially the Handbook of Graphs and Networks in People Analytics (McNulty 2022) and the Handbook of regression modeling in people analytics (McNulty 2021). We briefly touch on some of this in Chap. 2. In this category, we also include (Newman et al. 2011; Adams et al. 2020), and the seminal work of John Scott on social network analysis (Scott 2012), that can be considered a key bridging text between the quantitative aspects of network analysis and the social aspects of such networks.

Lastly, we should also mention that the accidentalilty of new groups is theoretically underpinned by the notion of **emergence** that can be traced back to seminal studies including (Holland 2000; Johnson 2002), linked to the cognate ideas of degrees of connections in a network (Watts 2004; Watts and Strogatz 1998; Watts 2014), and complexity (Mitchell 2009) and hidden order (Holland 1996).

Finally, it is worth mentioning that the ideas proposed in this book have a strong foundation in three seminal concepts developed in the past 10–20 years, especially for the examples that are provided in Chap. 2 on data sorting and group clustering, and the considerations presented in Chap. 5. First, Stephen Graham's work on software-sorted geographies (Graham 2014): an analysis of the role of software in shaping the social and geographical politics of inequality in advanced societies, is a key study focusing on the importance and role of code in the formation of politics that characterise inequality in today's societies. This study hinges on the idea that software-sorting approaches are to be related to a larger societal and cultural shift from "Keynesian to neoliberal service regimes" (Graham 2014:562). Albeit this study is almost 10 years old at the time of finishing this book, the main ideas are not only still valid, but have also been confirmed with the increased development of social media platforms in both size and overall reach over the last decade.

Second, Burrows and Gane's work on geodemographic classifications (Burrows and Gane 2006), provided a fundamental insight on the implication of software systems designed to sort out people and places, and how they connect social to physical (urban) spaces. Accidental Collectives partially originates from the idea that that social space is today to be considered in a new light, inextricably inseparable from geographical space, or what in architecture if defined as the built environment. This is because "the strength of geodemographic technologies is that they map social class in highly sophisticated ways" (Burrows and Gane 2006:808).

The final seminal idea is provided by Cynthia Dwork's notion of differential privacy (Dwork 2006): a way to algorithmically manage individual's privacy in data analysis). Dwork's studies examine the question of representation in datasets and data samples, where scientists need to find efficient and reliable ways to represent the characteristics of a given population while preserving a certain degree of privacy and other key features of individuals within the sample chosen.

A final note on the scientific domain of this book is that such study is indeed inter-disciplinary, for it crosses multiple subjects, from geography and geodemographics to social science, and from urban and design studies to computer and data science. One approach to organise the extensive literature that underpins the idea for Accidental Collectives would be to follow the established domains of human geography (e.g., Zook, Kitchin and Dodge, Marvin, Graham), information and communication technology (e.g., MacKenzie, Dwork, Dourish), and digital sociology (e.g.

Lupton, Bowker and Star, Boyd, Eubanks, Lampland). However, the intent of this book is to discuss the automatic formation of social groups as a byproduct of other priorities in the social media dynamics as a whole, where the different academic domains are intertwined.

## STRUCTURE OF THE BOOK

This book is organised into five chapters with the idea of taking the readers from a very general overview of code-driven data sorting to reflections on the implications of such classification of data (and people) into groups.

Chapter 2 illustrates how computers classify data into groups and build internal linkages. The chapter includes a very brief introduction of how machine learning (ML) techniques are typically used to classify people into groups and categories based on traits that are specified inside social platforms, which are then applied to people in different circumstances. Chapter 2 provides basic descriptions of some of the techniques typically used to group data into categories based on predetermined objectives (e.g., grouping users together by age range or region). The last section of the chapter goes into further detail about how machine learning techniques are used to create these groups in a reliable manner and with quantifiable metrics to gauge how effective the grouping is in relation to a certain objective.

Chapter 3 offers some insight into how communities are created around technology, and more particularly, how communities are created around the use of social media and online apps. In Chap. 3, I describe how individuals' inadvertent pursuit of self-affirmation and social recognition among a typically small circle of contacts leads to the emergence of enormous collectives. In order to gain their appreciation in the form of comments, likes, and shares from a small community of followers, each individual user of social media applications shares their everyday activities and accomplishments with them. These machine learning systems function as global organisations that span groups, contacts, and individuals, building larger, interconnected, and anonymous collectives that are not visible to or accessible to the individuals who compose them.

Chapter 4 shows how such intelligent systems function and quietly support our daily activities through three case studies. I discuss how the development of a set of a priori design criteria leads to the automatic formation of public spaces, common areas, and social locations where people gather, interact, and go about their daily lives in public. In particular,

Chap. 4 includes a range of projects where space is generated through algorithmic processes, from houses to care facilities, using machine learning techniques and AI-driven technologies in general. This part of the book explains how people's private and public lives are becoming increasingly mediated by software without their complete knowledge. A step-by-step illustration of some of the algorithmic logics that control how people move between their private and public lives and build software-driven collectives is presented in this chapter. This chapter explains some of the mechanisms that support the covert mechanisation of public life.

Finally, Chap. 5 examines how such automatically generated spaces relate to intentional and unintentional aspects of accidental collectives in terms of ethics, fairness, and social justice. This part of the book explores the ethical implications of the growth of accidental collectives by analysing how fair and equal they can and should be. The chapter focuses on the intricate relationships between AI and fairness. I discuss the difficulties that AI algorithms present in maintaining justice and avoiding bias in unintentional collectives. This section focuses on moral concerns about decision-making processes, data sources, and algorithmic transparency.

The final section of Chap. 5 examines the prospects for accidental collectives over the next 10 to 50 years. This section speculates on possible scenarios for the function of accidental collectives in society, spanning a spectrum from radical AI advancement to more practical and grounded advances. The chapter's conclusion makes the argument that when broad artificial intelligence becomes a reality, unintentional collective generation will become less common.

## MAIN POINTS TO TAKE AWAY

It is perhaps worth noting that Accidental Collectives starts from a technical perspective (how relevant algorithms and computational methods work in practice) to explain how concrete cases of automated social sorting operate. The book then considers how these automatedly generated collectives are (their qualitative characteristics) and what their possible evolution may be.

Unlike some of the interesting work mentioned above, which focuses on automating inequality, Accidental Collectives is not an account of how algorithms produce social inequality. It is rather an attempt to explore and explain one of the often-overlooked consequences of technological advancements in computing: the generation of spatial byproducts that, although relatively unintentional, have a significant impact on people's experiences and daily lives. The idea behind this work is to expose this phenomenon, explain how its mechanisms work, show some concrete applications to the digital and physical worlds, and leave it to the readers to formulate their own opinion on the matter. Throughout the book, the readers may find a number of ideas or takeaways, which I summarise here below.

## TRANSPARENCY AND PERCEPTION

This mostly emerges in Chap. 3, where we look at examples of how web applications and social media platforms put people into groups through predictions based on assumptions. We learn that such platforms are socially translucent instead of transparent. Users are in a position to perceive some characteristics of their network, but they never have a full picture. In fact, the relationship between the group of people directly available to each user and the full extent of their social network within the platform is unclear to the human user. Algorithms regulating the mechanisms of the platform have full access to the extent of the network, yet this is not their main focus. As we see in the following chapters, these algorithms are driven by certain objectives, which usually exclude social cohesion and transparency among users. As I explain in Chap. 3, social collectives of people are generated by mechanisms that have no direct interest in the characteristics of such groups or in their existence in a general sense. We therefore have the translucency that characterises the perception of individual users and their networks within the platform on the one hand. We then have on the other the accidentality of the generation of such networks that adds a degree of complexity to people's experiences of their own social networks.

## COMPLEXITY BEYOND HUMAN UNDERSTANDING

The idea of the social network achieving a level of complexity that users would struggle to deal with is further elaborated in Chap. 4, with concrete examples of space production. The readers will see how designers and developers generate intelligent systems with clear objectives to achieve and where human users have full control of inputs and outputs, yet are not able to comprehend the mechanisms underpinning the bulk of the computing process. I refer in Chap. 4 to the idea of blackbox, as a notion to explain that certain computational operations have a degree of complexity that surpasses any human comprehension. Both the developers of such intelligent systems and the final users of the platform have to make a leap of faith in the system, albeit from different positions. In fact, developers have a clear understanding of the step-by-step process of such mechanisms and the algorithms underpinning them. They know that, on a scientific level, established and tested methods will yield the desired results under the right conditions. On the other hand, social media users do not necessarily need to have any knowledge of the systems they use, as they simply trust the platform and its user experience features.

## Controlling Bias

Chapter 5 explores the presence and mitigation of bias in algorithmic processes. We easily learn that bias is actually a multifaceted inherited characteristic of algorithms and that there is no one way to address it to improve fairness and equality in the generation of social collectives. Instead, we have several helpful frameworks and guidelines. However, the main point here is that different approaches to bias mitigation successfully address specific aspects but not all. We learn here the importance of being open and transparent in the application of algorithms and choices through the process. I identify a mid-level of human-machine interaction (between scenarios where humans or machines are in utter control of the system) as the best dimension to improve the degree of transparency in the development of intelligent systems.

## Collectives Accidentality Is Inversely Proportional to General AI

Chapter 5 concludes with the idea that when general artificial intelligence becomes a reality, unintentional collective generation will become less common. Narrow AI is an intelligent system specialised in resolving specific problems, while a general artificial intelligence is able to address general problems, moving across knowledge domains. The end of this book suggests that it is more likely for a group to form accidentally if the artificial intelligence powering the clustering of users on a platform is narrow. On the other hand, this accidentality is less prevalent the more general the AI is. Collectives will become less accidental as AI gradually progresses from narrow to general.

## NOTES ON THE MAKING OF THIS BOOK

This book has been in the making for a few years. Chapter 2 includes studies and tests on data sorting and self-organising plans. Some of the studies included in Chap. 3 have been presented at the 16th Annual International Conference of the Architectural Humanities Research Association in Dundee, UK. Chapter 3 refers to Twitter in the discussion about social media. As the majority of published literature and references used in this book refer to the company name Twitter, I preferred to leave the old name (changed into X in 2023). As per March 2024, the main URL of the company is still twitter.com, although the name X is being used on the phone app and the website. At the time of writing it made sense to use Twitter, as readers may be more familiar with the old company name.

Chapter 4 includes three case studies that have been previously published and presented at conferences and seminars. Part of the literature review included in Chap. 2 has been developed in (Carta 2021) and further elaborated for this book. The first case study has been developed as a part of a project on automation and spatial improvement of care home design funded by Research England through the University of Hertfordshire (Carta et al. 2022). The second case study, Magnetising Floor Plans has been published in various forms in (Papadopoulou 2021; Papadopoulou et al. 2022).

Part of the third case study, Self-Organising Floor Plans in Care Homes has been published in (Carta et al. 2020). Some of the ideas discussed in Chap. 5 have been developed as a part of the UKRI funded network, NetworkPlus Not-Equal and the funded project MiniCode (Malizia et al. 2022), where we looked at ways to minimise algorithmic bias in collaborative decision-making using design fiction techniques.

Overall, the idea of Accidental Collective is supported by the past three to four years of studies on data-driven design, automation, intelligent spatial configuration, machine learning, artificial intelligence related to space and design, and ethical questions related to the interaction human-machine.

To produce this book, I have used the assistance of multiple AI-driven software. For example, in Chap. 2, I asked ChatGPT to do some of the heavy lifting for me in listing some of the algorithms for data sorting and clustering. I started by elaborating bullet-point lists of what I needed to be included in each section and created prompts for ChatGPT to elaborate and suggest some examples of practical use of some of the algorithms. I then paraphrased, checked and edited all texts and codes myself, but

ChatGPT provided me with a good start. All the code snippets are on the online repository created for this book, available at: https://github.com/seelca/Accidental-Collectives

## REFERENCES

Adams, J., Santos, T. and Williams, V.N. 2020. Strategies for collecting social network data. *The Oxford Handbook of Social Networks*, p. 119.

Amoore, L. 2011. Data derivatives: On the emergence of a security risk calculus for our times. *Theory, Culture & Society* 28(6), pp. 24–43.

Boyd, D. 2014. *It's complicated: The social lives of networked teens*. Yale University Press.

Burrows, R. and Gane, N. 2006. Geodemographics, software and class. *Sociology* 40(5), pp. 793–812.

Carta, S. (2021). Self-Organizing Floor Plans. *Harvard Data Science Review* 3(3). https://doi.org/10.1162/99608f92.e5f9a0c7.

Carta, S., St. Loe, S., Turchi, T. and Simon, J. 2020. Self-organising floor plans in care homes. *Sustainability* 12(11), p. 4393.

Carta, S., Turchi, T. and Papadopoulou, F. 2022. *AISLA, Analyse and Improve Spatial Layout of Care Homes*. Available at: https://www.herts.ac.uk/research/groups-and-units/arch/aisla.

Diakopoulos, N. and Koliska, M. 2017. Algorithmic transparency in the news media. *Digital Journalism* 5(7), pp. 809–828.

Dourish, P. 2022. *The stuff of bits: An essay on the materialities of information*. MIT Press.

Dwork, C. 2006. Differential privacy. In: *International colloquium on automata, languages, and programming*. Springer, pp. 1–12.

Ellison, N. and Boyd, D.M. 2013. Sociality through social network sites.

Elmer, G., Langlois, G. and Redden, J. 2015. *Compromised data: From social media to big data*. Bloomsbury Publishing USA.

Eubanks, V. 2018. *Automating inequality: How high-tech tools profile, police, and punish the poor*. St. Martin's Press.

Evans, J. and Ruane, S. 2019. *Data in society: Challenging statistics in an age of globalisation*. Policy Press.

Frischmann, B. and Selinger, E. 2018. *Re-engineering humanity*. Cambridge University Press.

Graham, S.D. 2014. Software-sorted Geographies:(2005). In: *The People, Place, and Space Reader*. Routledge, pp. 133–138.

Holland, J.H. 1996. *Hidden order: How adaptation builds complexity*. Addison Wesley Longman Publishing Co., Inc.

Holland, J.H. 2000. *Emergence: From chaos to order*. OUP Oxford.

Johnson, S. 2002. *Emergence: The connected lives of ants, brains, cities, and software*. Simon and Schuster.

Kearns, M. and Roth, A. 2019. *The ethical algorithm: The science of socially aware algorithm design*. Oxford University Press.

Kitchin, R. 2014. The data revolution: Big data, open data, data infrastructures and their consequences. *The Data Revolution*, pp. 1–240.

Kitchin, R. 2021. *Data lives: How data are made and shape our world*. Bristol University Press.

Lupton, D. 2015. *Digital sociology*. Routledge London.

Malizia, A., Carta, S., Turchi, T. and Crivellaro, C. 2022. MiniCoDe Workshops: Minimise Algorithmic Bias in Collaborative Decision Making with Design Fiction. In: *Proceedings of the Hybrid Human Artificial Intelligence Conference*.

McNulty, K. 2021. *Handbook of regression modeling in people analytics: with examples in R and Python*. CRC Press.

McNulty, K. 2022. *Handbook of graphs and networks in people analytics: with examples in R and python*. CRC Press.

Mitchell, M. 2009. *Complexity: A guided tour*. Oxford university press.

Newman, M., Barabási, A.-L. and Watts, D.J. 2011. *The structure and dynamics of networks*. Princeton university press.

Noble, S.U. 2018. Algorithms of oppression. In: *Algorithms of oppression*. New York university press.

O'neil, C. 2017. *Weapons of math destruction: How big data increases inequality and threatens democracy*. Crown.

Papadopoulou, F. 2021. *The development of a functional accommodation for female victims of sex trafficking: How its design will ensure the implementation of the programmes of assistance and the covering of their needs, leading to their practical and psychological support that will help them adjust and adapt back to 'normality'*. Hatfield: University of Hertfordshire.

Papadopoulou, F., Carta, S. and Owen, I., Wyn. 2022. Safe Houses: design principles, potentials and limitations. An analysis through data-driven approaches. In: *ATINER*. Athens, Greece: ATINER.

Persily, N. and Tucker, J.A. 2020. Social media and democracy: The state of the field, prospects for reform.

Scott, J. 2012. *What is social network analysis?* Bloomsbury Academic.

Turkle, S. 2005. *The second self: Computers and the human spirit*. Mit Press.

Watts, D. 2014. *Six degrees: the new science of networks*. Random House.

Watts, D.J. 2004. *Small worlds: the dynamics of networks between order and randomness*. Princeton university press.

Watts, D.J. and Strogatz, S.H. 1998. Collective dynamics of 'small-world' networks. *nature* 393(6684), pp. 440–442.

# CHAPTER 2

# Software Sorting

**Abstract** In this chapter, I look at how computers sort data into categories and link the data within them. Specifically, I outline how machine learning (ML) techniques tend to be used to create groups and categories. I also explore how these categories are then applied to people across a number of contexts, based on characteristics given on social media platforms. Through this, I hope to provide readers with an accessible overview of some of the techniques that computers use to create groups of people. This chapter starts by giving rudimentary explanations of the ML techniques used to separate data into categories based on specific goals (e.g., to combine all users within a specific age range or from the same geographical location). After exploring how data are sorted into the required groups, we shall see how connections between the data in each group are generated. The final part of the chapter illustrates how ML techniques are used to create these groups robustly and using quantifiable metrics, enabling us to assess just how effective a particular grouping is in meeting a specific goal.

**Keywords** Software sorting • Differential privacy • Classification • Data analytics • Data sorting • Grouping • Social collectives

© The Author(s), under exclusive license to Springer Nature Switzerland AG 2024
S. Carta, *How Computers Create Social Structures*,
https://doi.org/10.1007/978-3-031-62852-8_2

## DATA SORTING

Computers are ideally placed to organise information into different categories that can help us to create new ways of linking those pieces of information. At the same time, they can also enhance our understanding of existing connections. Once represented as data, information can be automatically classified, sorted, and interpreted using a number of well-established algorithms and methods; in turn, these methods can shed light on patterns that represent and often underpin human relationships. In this chapter, I shall focus in particular on how sorting can serve as a powerful mechanism for arranging data to meet strategic aims.

Perhaps one of the most intuitive ways to explain sorting is given by Bhargava (2016). Bhargava starts out from very simple data structures, lists, and arrays to illustrate how data can be arranged with specific goals or criteria in mind. Examples might include rearranging a list of songs alphabetically based on the names of their performers, or according to the number of times each song has been played. When we rearrange data for a specific purpose, we select aspects of the data we have to hand in order to understand those data better or, as is perhaps more often the case, to produce new knowledge. It is, therefore, no surprise that sorting is a fundamental operation in computer science, where it has a wide range of uses, in everything from search engines and data mining to ML and data analysis.

In this section, I outline some of the methods that are most frequently used to sort data into categories and I illustrate how computers make use of them. The algorithms I introduce here are intended to form a solid basis for the reader. This basis will help us as we explore how some of these methods work in practice in subsequent chapters.

When categorising data, our first, essential step is to identify the attributes that will determine our categories. If, for example, we want to categorise a collection of books methodically, we need to consider a number of criteria, including authors' names, genres, publication dates, and other relevant features that are intrinsic to the collection. Our choice of particular attributes to be used when sorting data is contingent on both the characteristics of the data themselves and our overarching goal in categorising them. And if we are careful and systematic when selecting our criteria, we can arrange and classify the data in the optimal way to meet our objective.

A variety of algorithms are used widely for sorting tasks. These algorithms include **insertion sort, selection sort, bubble sort**, and **quicksort**, and the choice of which one to use largely depends on the type and volume of data to be classified, as well as the degree of computational efficiency required for the sorting task in question.

The **insertion sort** algorithm sorts data by iterating through an array of elements and carefully comparing each item in that array with the components that precede it. It involves inserting one element into the array at a point that aligns with that element's new location, as determined by the sorting process. The sorting process determines an element's new location by comparing the components inside the array against one another. As it progresses, the algorithm systematically moves elements within the array to the right iteratively until it identifies the appropriate point at which the new element should be added. Owing to its straightforward implementation and its efficiency, the insertion sort algorithm is particularly effective when handling small data sets. But its speed and efficiency suffer when it is applied to larger data sets that require it to perform far more comparisons and shifts.

The pseudocode below gives the formal description for the insertion sort algorithm:

```
for i = 1 to n-1
    j = i
    while j > 0 and A[j-1] > A[j]
        swap A[j] and A[j-1]
        j = j - 1
```

Code snippet 2.1 Loop for insertion sort algorithm

Here, $A$ is the array that needs to be sorted, $n$ denotes the number of elements in the array, and $i$ and $j$, both of which are integers, are variables used for iteration and comparison. The outer loop of this pseudocode iterates through each element of the array, beginning with the second element ($i=1$) and continuing until it reaches the subsequent element, as defined by the condition $i < n-1$. The current element is compared with the elements that precede it ($j > 0$) in the inner loop. If the preceding element is greater than the current element, then the two elements are switched around ($A[j-1] > A[j]$). This process continues until it has found the appropriate location for the current element.

If we want to evaluate the efficiency of the insertion sort algorithm, we can use the following equation:

$$C(n) = 1 + 2 + 3 + \ldots + (n-1) = (n^2 - n)/2$$

This equation calculates the sum of the first $n$-1 integers, giving the total number of comparisons performed by the procedure while sorting the data. This value, in turn, indicates how complex the algorithm is. The insertion sort algorithm's temporal complexity is represented in computational notation as $O(n^2)$, showing that this temporal complexity is proportional to the square root of the number of elements in the array. For a more detailed discussion of big O notation and computation times, see Bae (2019:1–11).

Like the insertion sort algorithm, **selection sort** is another basic sorting method. It works by repeatedly selecting the element with the lowest frequency of all of the unsorted elements and transferring it to the beginning of the sorted array, which contains nothing at the start. It repeats this procedure iteratively until the whole array has been sorted. Again, like the insertion sort algorithm, selection sort is relatively efficient when managing relatively small datasets. Its efficiency noticeably decreases, however, when it is employed on larger sets and has to work its way through much greater numbers of iterations and comparisons.

The **bubble sort** algorithm works by constantly swapping the positions of elements that are near one another but in the incorrect order. It starts with the first two elements in the unsorted array and carries on swapping elements until the array is sorted. While bubble sort is fairly easy to understand and to put into practice, it is too slow and inefficient to be used for large datasets.

The **quicksort** algorithm proceeds on a basis of divide-and-conquer, thus making it a more efficient sorting method. It is a multi-step process that begins by dividing the unsorted array into two separate sub-arrays. A pivot element is selected and placed in one of the sub-arrays, while the other sub-array contains elements greater than or equal to this pivot element. The technique then iteratively sorts each of the sub-arrays by sorting and combining smaller sub-arrays to form larger sub-arrays, doing so repeatedly until the entire array is sorted.

The quicksort algorithm can be represented using the following pseudocode:

```
#quick sort and partition
quicksort(A, lo, hi):
    if lo < hi:
        pivot_index = partition(A, lo, hi)
        quicksort(A, lo, pivot_index-1)
        quicksort(A, pivot_index+1, hi)

partition(A, lo, hi):
    pivot = A[hi]
    i = lo
    for j = lo to hi-1:
        if A[j] < pivot:
            swap A[i] and A[j]
            i = i + 1
    swap A[i] and A[hi]
    return i
```

Code snippet 2.2 Quicksort algorithm

This pseudocode contains a number of important components. Alongside the array $A$, which is to be sorted, we have $lo$ and $hi$, which are the indices designating the boundaries of the sub-arrays. It also employs the integer variables $pivot\_index$, $i$, and $j$ in order to compare and partition the data. The pseudocode sets out a repeated sorting operation within the range of the supplied sub-array. It starts with the partition function, which divides the array into two sub-arrays, one of which contains elements that are less than or equal to the pivot value and the other of which contains elements that are greater than or equal to the pivot value. Once the data have been partitioned along these lines, the quicksort method sorts the two resulting sub-arrays by repeating this process using updated sub-array boundaries. This process continues until all sub-arrays have been sorted and all elements in the entire array $A$ have been placed into the right order. Quicksort is a particularly effective method for sorting large datasets because it can rapidly divide data into smaller subsets that can be sorted in turn. And, as a result, it tends to be used widely.

We can measure the efficiency of the quicksort algorithm using the equation:

$$C(n) = n^* \log(n)$$

This gives us a rough estimate of how many comparisons the algorithm makes on average when the pivot element has been chosen at random. Its temporal complexity is expressed in computational terms as O(n*log n), indicating that the amount of time quicksort requires to sort an array of *n* items increases according to the logarithm of the array's size. Given its efficiency in working with relatively small sets, therefore, quicksort has become especially popular as a method for simple sorting applications.

**Merge sort**, like quicksort, takes a divide-and-conquer approach to data. It partitions an unsorted array into smaller sub-arrays, which are then repeatedly combined into larger, sorted sub-arrays. This process continues until all elements are combined in a single, sorted array.

While the sorting methods outlined above are the fundamental approaches used in computing, a large number of further sorting approaches have been developed for use with particular kinds of data. The **radix sort** algorithm, for example, is designed to sort integers, while **bucket sort** works by sorting data into buckets.

As we can see, the first step in gaining deeper insights from data requires that these data are sorted into specific categories or sub-arrays. Once this has taken place, we can undertake further analysis and obtain more detailed insights about the data we have to hand. We might, for example, perform statistical analysis on the data within these categories to learn more about the properties of each set and how the elements within them are distributed. ML is particularly useful in this regard, as it enables us to group newly collected data into pre-existing categories, even providing the basis from which we can predict or forecast future trends. In the following section, therefore, I focus on the operations that can be performed using these newly sorted data; and I discuss in particular how we can use ML to identify links across different sets of data.

## LINKING FEATURES ACROSS CATEGORIES

Algorithms play an essential role in generating patterns and finding connections between different categories in datasets. Data mining, for example, studies large datasets and extracts usable information and insights before finding patterns and linkages within the data. This section looks at how algorithms do so.

In order to be ready for analysis, a dataset must first be pre-processed. This comprises correcting any flaws or inconsistencies in the data as well as translating the data into a format that can be utilised by the algorithms without missing anything out. We tend to call these stages "data cleaning" and "data wrangling". Once the data are clean, by which we mean that they are all in the same format and ready to be processed, we can begin to apply algorithms that will extract patterns and identify relationships between the different categories.

The **Apriori** algorithm is one of the most commonly used algorithms employed for data mining. It generates association rules by first locating itemsets (sets of items that are considered together) that occur frequently within in a dataset and then generating rules based on these itemsets. If we were looking at a dataset of client purchases, for instance, we may want to use the Apriori algorithm to determine which products are usually purchased in combination. We can then use this information to build association rules, such as "if a customer buys item A, they are likely to buy item B as well."

To create candidate itemsets, the Apriori algorithm undertakes a breadth-first search. This looks through a data tree for specific nodes with specific characteristics. Once the tree has been searched, the Apriori algorithm *prunes* the search space by no longer considering those itemsets that do not meet its minimum support threshold (pre-established characteristics).

The Apriori algorithm is formally described using the pseudo-code below:

```
#A priori algorithm
Apriori(D, minsup):
    L1 = frequent_1_itemsets(D, minsup)
    L = [L1]
    k = 2
    while L[k-2] != []:
        Ck = generate_candidates(L[k-2])
        Lk = frequent_k_itemsets(Ck, D, minsup)
        L.append(Lk)
        k = k + 1
    return union(L)
```

Code snippet 2.3 Apriori algorithm

Here, $D$ refers to the dataset to be studied, *minsup* is the minimum support threshold, $Lk$ refers to the list of frequent k-itemsets, and $Ck$ to the list of candidate k-itemsets.

The Apriori algorithm begins by locating the frequent 1-itemsets the dataset $D$. It does this by using the *frequent_1_itemsets* function, which provides a list of the recurring itemsets that fulfil the minimal support criterion. Following this, the list, $L$, is initialised with the value $L1$, and the $k$ variable is set to the value 2, specifying that each itemset that is to be formed will have two items in it. The algorithm then initiates a loop that keeps running until it can find no more common itemsets. The *generate_candidates* function uses the itemsets found in *L[k-2]* to produce candidate itemsets that have the length *k*. The *frequent_k_itemsets* function then searches through all of the candidate itemsets for those containing the most frequent occurrences of the minimal support threshold *minsup*. After that, the frequently occurring itemsets discovered in *Lk* are added to the *L* list, and *k* is increased by one so that the process can continue, this time looking at itemsets that are one item larger. This loop continues until it can find no more frequent itemsets, and the union function returns return a list containing all of the recurring itemsets that were discovered.

We can use the following expression to evaluate the Apriori algorithm's efficacy with regard to generating candidate itemsets:

$$C(n) = (n \text{ choose } k) = n! / ((n-k)! k!)$$

This equation shows the number of possible combinations of *k* items that can be found in a dataset consisting of *n* items. The time it takes the algorithm to identify recurring itemsets increases in proportion to the square root of the number of items that are contained in the dataset. This temporal complexity is defined as $O(n^2)$.

The **Decision Tree** algorithm provides a reliable approach to categorisation analysis. If we want to maximise its potential, we first need to divide the dataset into a number of subsets depending on the values of particular attributes. The algorithm then builds a tree-like structure, showing the connections between these attributes and their related categories. Imagine, for instance, that we have collected certain characteristics of cars. Using the decision-tree technique, we could classify cars as sports or family vehicles based on parameters like horsepower and passenger capacity. With the data now grouped into a number of clusters, we can take another look at the findings, investigating patterns, correlations, and anomalies in them, and draw some new conclusions. If we turn back to our dataset of consumer purchases, for example, we might be able to identify seasonal variations in requests, commonly purchased items, and customer preferences for specific products. Data mining techniques reveal these patterns, but can also use them as the basis for offering forecasts and tailored suggestions. Such forecasts might include pointing to potential connections or offering personalised recommendations for a particular user, all drawing on a dataset of user preferences and, in the process, increasing the usefulness of the data. In the next section, we shall look at ML techniques make these things possible.

## How Machine Learning (ML) Techniques Are Used to Create Groups

ML techniques are increasingly being used to establish groupings and identify trends in data. In this section, I focus specifically on how social media makes use of ML techniques to these ends. We can apply the methods outlined above to analyse data from pretty much all social media platforms. Doing so would enable us to recognise patterns, develop suggestions strategies, and, ultimately, generate greater usage among subscribers, thus making these platforms more successful. Social media platforms produce enormous volumes of data, including everything from user-generated content and comments posted by other users to likes and shares. According to James (2022), in 2022 internet users shared 1.7 million content items on Facebook, conducted 5.9 million searches on Google, sent 231.4 million emails, shared 347,000 tweets on Twitter, and posted 66,000 images on Instagram every minute. The use of these platforms has grown steadily over the past decade, and we have every reason to believe that they will continue to do so. Such enormous quantities of rich data are very helpful when it comes to analysing and mining data, because large datasets usually correlate with more accurate results and predictions. And sophisticated ML techniques are extremely useful if we want to gain insights into users' behaviours and preferences. Not only do they help us to understand social media usage in the past; they also enable us to make predictions about future trends.

Broadly speaking, we can identify two main types of ML techniques, based on the form of learning use in them. On the one hand is **supervised learning**, which is one of the most common forms of ML used in social media. Supervised learning starts with an algorithm that has been trained using a labelled dataset. As a result of this training, the appropriate output for each input in this dataset is already known; so, for example, a picture of a cat is equal to a cat and therefore goes into the "cat" category when sorted. Through this training process, the algorithm learns from its experiences both to recognise patterns and how to generate predictions based on them. **Unsupervised learning**, on the other hand, starts out not with labelled data; instead, the algorithm has to discover patterns and structures within the data on its own.

Alongside these two main types of ML learning, it is also worth mentioning **reinforcement learning**. Here, the algorithm learns through a process of trial and error, receiving rewards or punishments for specific actions. Reinforcement learning is a powerful tool for optimising the outreach of social media and increasing user engagement. In this context, a reinforcement learning algorithm can, for example, be used to determine when distributing customised material to a particular audience will have the maximum effect and impact. These algorithms evaluate and respond to user replies and engagement data to generate optimal posting schedules, forms of content, and interaction methods. And in so doing, they learn how to increase the reach and impact of content on social media.

## How Recommendation Works

A good example of supervised learning in action in the context of social media is the implementation of **recommendation algorithms**. Social media platforms frequently use these algorithms when recommending that users follow specific accounts or engage with new material. Broadly speaking, recommendation algorithms analyse users' actions and interests so that social media platforms can predict their likely future interactions. Users therefore receive personalised suggestions based on their previous interactions with goods, such as movies, music, books, or products. Recommendation algorithms take many forms, but two that are worth particular attention here are **collaborative filtering** and **content-based filtering**.

The **collaborative filtering** algorithm analyses users' social media activities and looks for patterns in the interactions between users and goods. It then draws on the behaviour of other users with similar profiles as the main source for making recommendations. **Content-based filtering**, however, examines the characteristics of the items that are filtered and looks for recurring patterns in the contents of these items. The algorithm then uses this information alongside data regarding users' previous activities to recommend goods and services that are comparable to the recommended product.

The effectiveness and accuracy of these recommendation algorithms tends to be measured using a number of metrics, including **precision**, **recall**, **F1-score**, and **Mean Absolute Error (MAE)**. The **precision** metric determines what percentage of items recommended to the user are actually relevant, whereas **recall** determines what percentage of relevant items have actually been recommended to the user. The **F1-score** is the harmonic mean of the precision and recall scores, providing a more balanced assessment of the recommendation algorithm's performance. The

**MAE** metric measures the algorithm's accuracy at predicting user preferences, doing so by measuring the difference between the items' predicted and actual ratings. These four metrics are more expressed formally by the following equations:

$$Precision = \frac{(\text{number of relevant recommended items})}{(\text{total number of recommended items})}$$

$$Recall = \frac{(\text{number of relevant recommended items})}{(\text{total number of relevant items})}$$

$$F1-score = 2\frac{(\text{precision} * \text{recall})}{(\text{precision} + \text{recall})}$$

$$MAE = \frac{\sum_{i=1}^{n}(\text{predicted rate} - \text{actual rating})}{n}$$

Here, $n$ represents the total number of items and the *sum function* compiles the totals for each individual item in the dataset.

In summary, recommendation algorithms present users with personalised recommendations based on their previous experiences of and interactions with the platform in question.

When looking for patterns in how users have interacted with content (or other users) on a platform, however, we may encounter inconsistencies. And these need to be accounted for. A good way in which we can address these inconsistencies is to use an **anomaly detection** algorithm. As we cannot know what the anomaly is in advance (nor do we know the category in which such data need to go), the anomaly detection algorithm deploys a form of unsupervised learning. It identifies suspicious or unusual conduct on social media platforms, finding activities that do not fit with the behaviour that has been predicted. These activities might indicate the presence of false profiles or accounts that engage in spamming, trolling, and other disruptive or even illegal activities.

In a broader sense, however, anomaly detection algorithms analyse datasets by spotting unusual occurrences or unanticipated patterns. Although both supervised and unsupervised approaches to anomaly detection are possible, here I want to concentrate on the unsupervised mode, because it is particularly relevant to this study. Unsupervised anomaly detection does not rely on past knowledge; nor does it work with data that have been labelled as normal or anomalous. Instead, it works on the assumption that the majority of the data are normal and recognises patterns that are statistically unlikely based on this premise. It performs best when the underlying anomalies are unknown or where very few data are labelled. This makes unsupervised anomaly detection algorithms extremely promising for a variety of applications. Through their combination of statistical and ML techniques to automatically discover outliers and deviations, they can provide important insights into unforeseen events or inconsistencies inside datasets.

Much like the recommendation algorithms discussed above, we can measure the effectiveness and reliance of anomaly detection algorithms using **precision, recall**, and the **F1-score**; added to these, however, we can use the area under the receiver operating characteristic curve (**AUC-ROC**). The ROC curve is a graphical representation of a binary classifier system's diagnostic capabilities when the threshold for discriminating varies.

While the **recall** metric counts the proportion of genuine positives that are accurately classified as positives, the **precision** metric calculates the percentage of true positives relative to the total number of predicted positives. As the harmonic mean of precision and recall, the **F1-score** offers a more balanced picture of the anomaly detection algorithm's performance. The **AUC-ROC** measures how different categorisation criteria affect the ratio of true positives to false positives. While a random classifier would receive a score of 0.5 on the AUC-ROC scale, a perfect classifier would receive a score of 1.0. These metrics are formalised using the following equations:

$$Precision = \frac{(\text{true positives})}{(\text{true positives} + \text{false positives})}$$

$$Recall = \frac{(\text{true positives})}{(\text{true positives} + \text{false negatives})}$$

$$F1 - score = 2\frac{(\text{precision} * \text{recall})}{(\text{precision} + \text{recall})}$$

$$AUC - ROC = \textit{area under the receiver operating characteristic curve}$$

In short, anomaly detection algorithms are essential if we want to identify unusual occurrences or patterns within our data. On a very rudimentary level, these methods generally account for a balance between expected and usual behaviours within patterns observed on social media platforms.

Above, I have outlined how content-filtering and collaborative filtering algorithms work to offer recommendations. We have also seen how algorithms can detect anomalies within patterns. In the following section, therefore, we shall turn to how ML techniques can be used to infer new information from social media data.

### *Reading Between the Lines*

Once we have used ML techniques to identify patterns within social media users' interactions, we can take a look at the content of those interactions. At this point, we might, for example, analyse the semantics of the communication exchanged within the platform. If we wanted to study the emotional tone and context of user-generated information, then we would use **sentiment analysis**. This is a particularly good tool for analysing attitudes and sentiments expressed in user communication and can assess whether a positive, negative, or neutral attitude is being conveyed. Sentiment analysis uses a wide range of methods to determine the meanings of text, from simple rule-based approaches that rely on dictionaries to associate words with particular emotions, to more sophisticated approaches based on ML and natural language processing (NLP). We can analyse the following sentences using the simplest rule-based approach:

> I dislike like the updated layout of the platform. It's hard to use and confusing.

From this, we can easily infer that the text contains negative elements. Sentiment analysis would identify that the words "dislike," "hard," and "confusing" have negative connotations within the context of the message. More sophisticated approaches, including those that involve NLP, would analyse the text using statistical models and algorithms in order to identify patterns connected with various emotions. These models are trained on large datasets of text that have already been labelled with sentiments, and they can be used to categorise datasets of text based on the patterns that were detected during training. Our model might, for example, interpret the following message as having positive connotations:

> I just had the best meal of my life at this new restaurant. The food was amazing and the service was outstanding!

The model may recognise a pattern in the use of the positive adjectives "best," "amazing," and "outstanding," and therefore determine that the overall statement is positive. It may also consider where these words occur within sentences and the larger context of the text as a whole.

Figure 2.1 illustrates how specific words are quantitatively assessed using sentiment analysis. Here, we see that it looks at the number of times

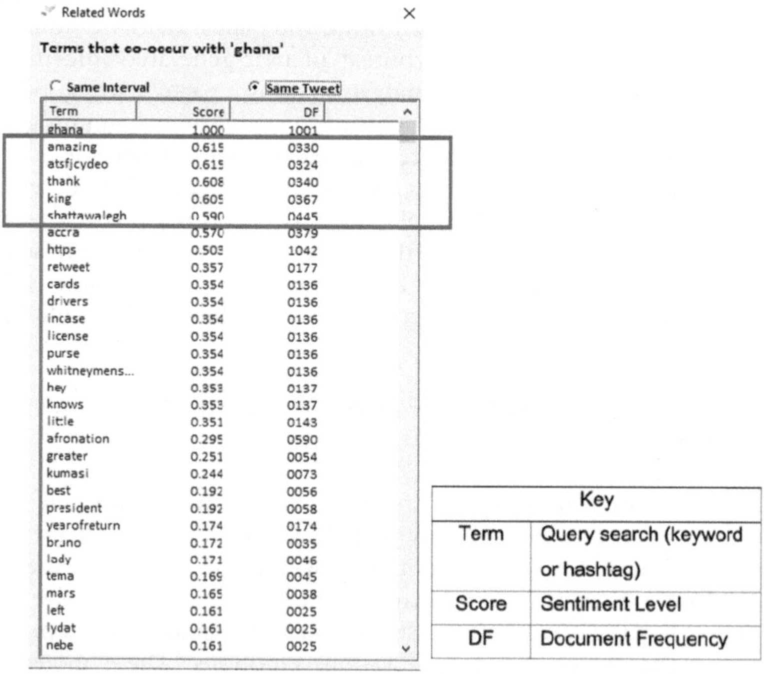
Related Words

Terms that co-occur with 'ghana'

Same Interval | Same Tweet

| Term | Score | DF |
|---|---|---|
| ghana | 1.000 | 1001 |
| amazing | 0.615 | 0330 |
| atsfjcydeo | 0.615 | 0324 |
| thank | 0.608 | 0340 |
| king | 0.605 | 0367 |
| shattawalegh | 0.590 | 0445 |
| accra | 0.570 | 0379 |
| https | 0.503 | 1042 |
| retweet | 0.357 | 0177 |
| cards | 0.354 | 0136 |
| drivers | 0.354 | 0136 |
| incase | 0.354 | 0136 |
| license | 0.354 | 0136 |
| purse | 0.354 | 0136 |
| whitneymens... | 0.354 | 0136 |
| hey | 0.353 | 0137 |
| knows | 0.353 | 0137 |
| little | 0.351 | 0143 |
| afronation | 0.295 | 0590 |
| greater | 0.251 | 0054 |
| kumasi | 0.244 | 0073 |
| best | 0.192 | 0056 |
| president | 0.192 | 0058 |
| yearofreturn | 0.174 | 0174 |
| bruno | 0.172 | 0035 |
| lady | 0.171 | 0046 |
| tema | 0.169 | 0045 |
| mars | 0.165 | 0038 |
| left | 0.161 | 0025 |
| lydat | 0.161 | 0025 |
| nebe | 0.161 | 0025 |

| Key | |
|---|---|
| Term | Query search (keyword or hashtag) |
| Score | Sentiment Level |
| DF | Document Frequency |

**Fig. 2.1** Example of sentiment analysis on Twitter posts. Source: Onafuye (2021:215)

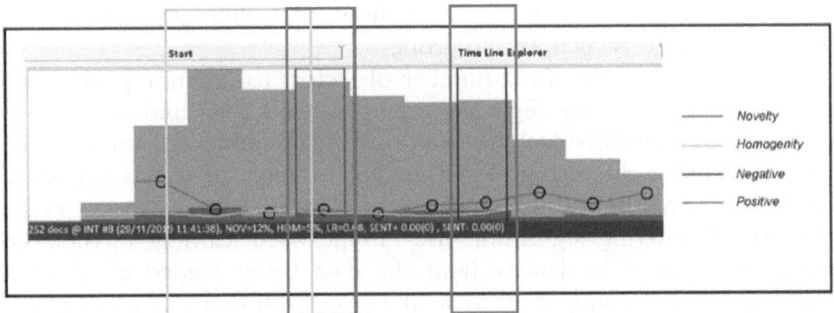

**Fig. 2.2** Example of sentiment analysis on a timeline. Source: Onafuye (2021:216)

these words appear in the message and their overall sentiment level score within the entire message or thread.

We can conduct a more sophisticated analysis by observing trends in and the homogeneity of sentiment across an entire conversation or thread. Figure 2.2, for example, illustrates how positive and negative elements, along with novelty and homogeneity, develop throughout the course of a series of messages.

It is also worth noting that we can perform sentiment analysis using hybrid methods. In some cases, we might produce more accurate results if we combine rule-based approaches with ML techniques. A hybrid tool for sentiment analysis might, for example, start by using a rule-based methodology to determine the tone of a piece of text. Following that, the tool might use an ML algorithm that allows it to fine-tune its categorisation based on the specific context of the text in question.

One of the most important recent studies of NLP and text-based analysis is found in the work of Vaswani et al. (2017), who emphasise the need for attention to context and self-attention. Their research proposes "a new simple network architecture, the Transformer, based solely on attention mechanisms, dispensing with recurrence and convolutions entirely" (Vaswani et al. 2017:1).

### *Clustering People into Groups*

Now that we have seen how patterns are identified and how more information about both the context and the overall tone of the communication can be gathered, we can start to look at how ML techniques use this information to categorise people into groups.

ML approaches can use a number of factors to organise people into groups or clusters. They might, for example, select characteristics or types of behaviour they individuals have or display in common with one another. We could use a **clustering** algorithm, for example, to categorise social media users according to the preferences or interests they express on the platform. Clustering algorithms use unsupervised learning methods to recognise patterns in data without the data being tagged in advance. Missing out the tagging stage generally saves both time and effort in the analysis process. Clustering algorithms set out to group data points into clusters based on the similarities between those individual pieces of data. They do this while simultaneously increasing the distance between the various groups in order to emphasise the mutual differences characterising each group. A wide variety of clustering algorithms are in widespread use, and the most popular among them are **K-means**, **hierarchical clustering**, and **DBSCAN** (Density-Based Spatial Clustering of Applications with Noise).

The **K-means clustering** algorithm is a very popular method for clustering analysis and data mining. It works by partitioning a dataset into a pre-determined number $(K)$ of clusters. It starts by randomly assigning each piece of data to one of the $K$ clusters before computing the mean of each cluster and placing each datapoint into the cluster whose mean is its closest match. This procedure is carried out repeatedly until the cluster allocations are consistent with one another and each datapoint has been assigned to a cluster.

Let us take, for example, a dataset that represents the interests of people using a social media platform. Here, each user is represented as a vector of feature values. We can use K-means to categorise people according to their shared interests and classify them into clusters. If we wanted to categorise the users into four distinct groups (clusters) using K-means, two things would happen. Firstly, the algorithm would randomly allocate each user to one of the four clusters. Then, it would determine the mean for each cluster and place each user in the group whose mean is closest to the average value of each cluster, doing so repeatedly until the cluster allocations are consistent. A simple implementation of a K-means clustering algorithm can be illustrated as follows:

```python
import numpy as np
import matplotlib.pyplot as plt
from sklearn.datasets import make_blobs

# Generate random data points using scikit-learn's make_blobs
data, _ = make_blobs(n_samples=300, centers=4, random_state=42)

# Define the number of clusters (K)
K = 4

# Initialize centroids randomly from the data points
centroids = data[np.random.choice(data.shape[0], K, replace=False)]

def kmeans(data, centroids, max_iters=100):
    for _ in range(max_iters):
        # Assign each data point to the nearest centroid
        labels = np.argmin(np.linalg.norm(data[:, np.newaxis] - centroids,
axis=2), axis=1)

        # Update centroids by computing the mean of each cluster
        new_centroids = np.array([data[labels == k].mean(axis=0) for k in
range(K)])

        # Check for convergence
        if np.all(centroids == new_centroids):
            break

        centroids = new_centroids

    return centroids, labels

# Run K-means algorithm
final_centroids, cluster_labels = kmeans(data, centroids)

# Plot the results
plt.figure(figsize=(8, 6))
plt.scatter(data[:, 0], data[:, 1], c=cluster_labels, cmap='viridis',
edgecolors='k')
plt.scatter(final_centroids[:, 0], final_centroids[:, 1], c='red', marker='X',
s=200, label='Centroids')
plt.legend()
plt.title('K-means Clustering')
plt.xlabel('Feature 1')
plt.ylabel('Feature 2')
plt.show()
```

Code snippet 2.4 K-means algorithm

The first step in this illustration uses the *make_blobs* function from scikit-learn to generate a few random datapoints. The centroids are then initialised based on a random selection of the datapoints. The K-means method is executed by the *kmeans* function, which does so in an iterative manner until either convergence or the maximum number of iterations (*max_iters*) is achieved. This function is used to obtain the final centroids as well as the labels for the clusters.

We then plot the datapoints using the labels that correspond to the clusters and highlight the centroids on the plot with red crosses (Fig. 2.3).

As we can see, the K-means algorithm is sensitive to the initial positions of the centroids, as well as to the number of clusters decided on at the outset. Depending on the distribution of the data, the algorithm may converge on a variety of distinct local minima. In general, if we are trying to find clustering solutions, it can be helpful to perform a number of runs, each one beginning with different centroids.

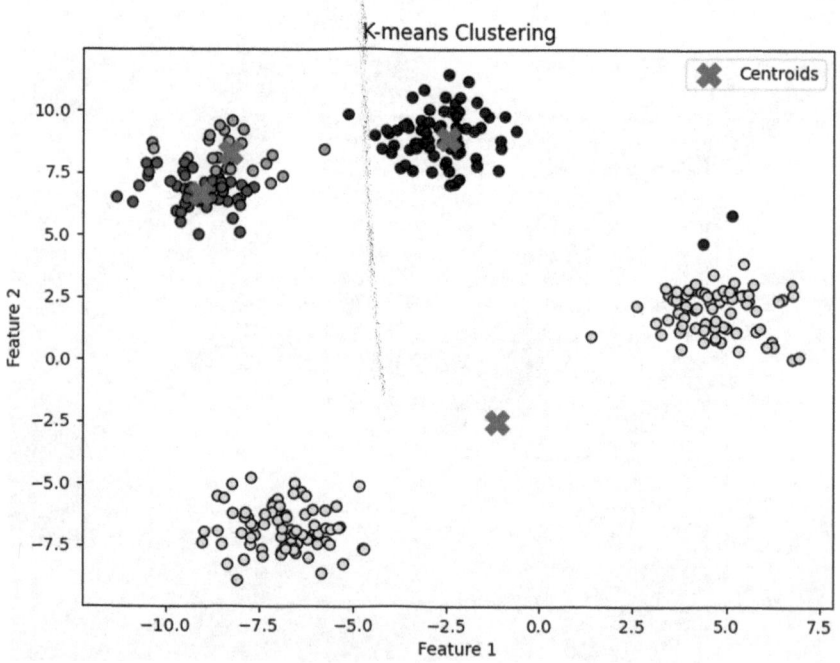

**Fig. 2.3**   Clusters generated using K-means

Another widely used clustering algorithm is **hierarchical clustering.** This algorithm organises data by creating a hierarchy of clusters. It starts by analysing each datapoint as if each were a cluster in its own right. It then repeatedly merges the clusters that are closest to one another until all of the datapoints are included in the same cluster. To see how this works, we can turn back to the datasets used in our previous example (illustrating a K-means clustering algorithm) in which our data comprised users' interests as shared on a given social media platform. In this case, hierarchical clustering categorises users into groups according to the topics that interest them most. The algorithm would initially generate as many clusters as the number of users in the dataset. It would then iteratively merge these clusters by using their means to determine their position until all users are included in a single cluster. This agglomerative clustering process is illustrated in Fig. 2.4.

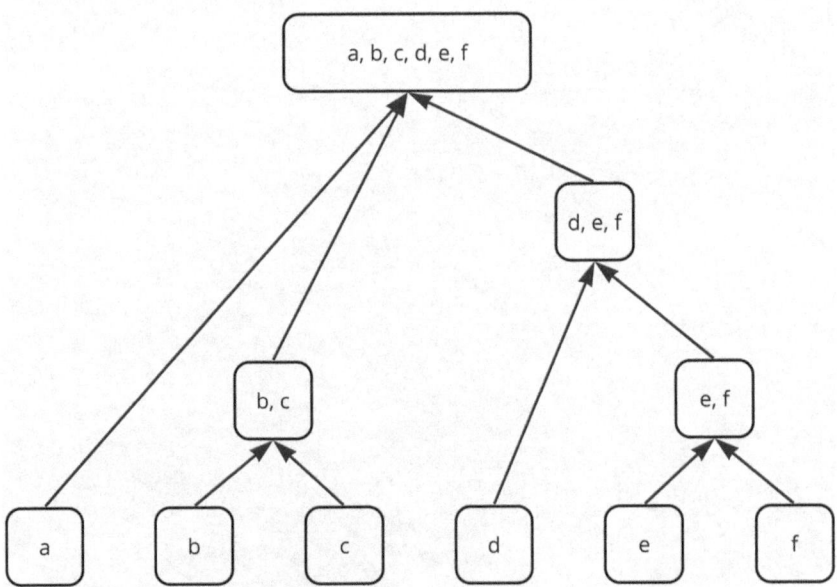

**Fig. 2.4**  Hierarchical clustering and iterative merging

The third method that I shall outline here is **DBSCAN** (Density-Based Spatial Clustering of Applications with Noise). This algorithm works by grouping together those datapoints that are closely packed together while excluding any datapoints that are isolated (outliers). This method first establishes a metaphorical neighbourhood around each datapoint. It then collects all of the datapoints that are located in neighbourhoods that have a high population density. DBSCAN categorises datapoints into subsets according to specific features listed in the database, such as, for example, geographical location, people who went to the same school, or people with similar political views. Once a neighbourhood of interests is built around each person, DBSCAN groups together those social media users who are part of a dense neighbourhood and excludes the outliers. DBSCAN therefore enables us to generate reliable clusters of social media users based on their interests (or any other factor we might be interested in). A DBSCAN might take the following form:

```python
from sklearn.cluster import DBSCAN
from sklearn.datasets import make_blobs
import matplotlib.pyplot as plt

# Generate some sample data
X, y = make_blobs(n_samples=1000, centers=3, n_features=2, random_state=42)

# Create a DBSCAN object with eps=0.5 and min_samples=5
dbscan = DBSCAN(eps=0.5, min_samples=5)

# Fit the data to the DBSCAN model
dbscan.fit(X)

# Extract the labels and the number of clusters
labels = dbscan.labels_
n_clusters = len(set(labels)) - (1 if -1 in labels else 0)

# Plot the clusters
plt.figure(figsize=(10, 8))
plt.scatter(X[:, 0], X[:, 1], c=labels, cmap='viridis')
plt.title(f"DBSCAN clustering with {n_clusters} clusters")
plt.xlabel("Feature 1")
plt.ylabel("Feature 2")
plt.show()
```

Code snippet 2.5 DBSCAN algorithm

In the above piece of code, we start by generating sample data using the *make_blobs()* function, located in the sklearn.datasets package. We then create a DBSCAN object with an *eps* value of 0.5 and a *min_samples* value of 5. These settings determine the algorithm's sensitivity to noise as well as its ability to identify clusters of varying sizes and shapes. After that, we used the *fit()* technique to adjust the data so that it is more compatible with the DBSCAN model. The cluster labels that were applied to each individual datapoint are stored in the *labels_attribute* of the model object. The number −1 denotes noise points that do not belong to any particular cluster. Once the algorithm has run its course, we depict the clusters with the help of the *matplotlib.pyplot.scatter()* function. The *c* parameter is set to labels. This means that the data points can be colour-coded to show which clusters they belong to. Figure 2.5 illustrates how all datapoints (users) have been sorted into three distinct clusters.

### *Computing Assumptions*

Clustering methods present us with powerful tools for grouping social media users into categories based on specific features of interest. We would use clustering, for example, if we wanted to locate all users of a single platform who are potentially interested in a particular political party, cultural event, or new product to be launched on the market. These methods are pretty straightforward to use when we have all the data we could possibly want about social media users and their preferences. That said, social media users rarely input all of the information we might need to perform a particular analysis or apply a predictive algorithm. When that is the case, we have to work with assumptions, using these to fill any gaps between datapoints. We can choose from a number of effective methods for filling gaps in our data, but in the following I shall focus on graph-based approaches, and specifically feature propagation, which is an uncomplicated and efficient method for inferring data (Bronstein 2022).

To put this in simple terms, let us imagine a social network, modelled as a graph. Here, the nodes represent the users, the edges represent the connections between them, and every node has a number of features (or attributes) that characterise their profile, including age, location,

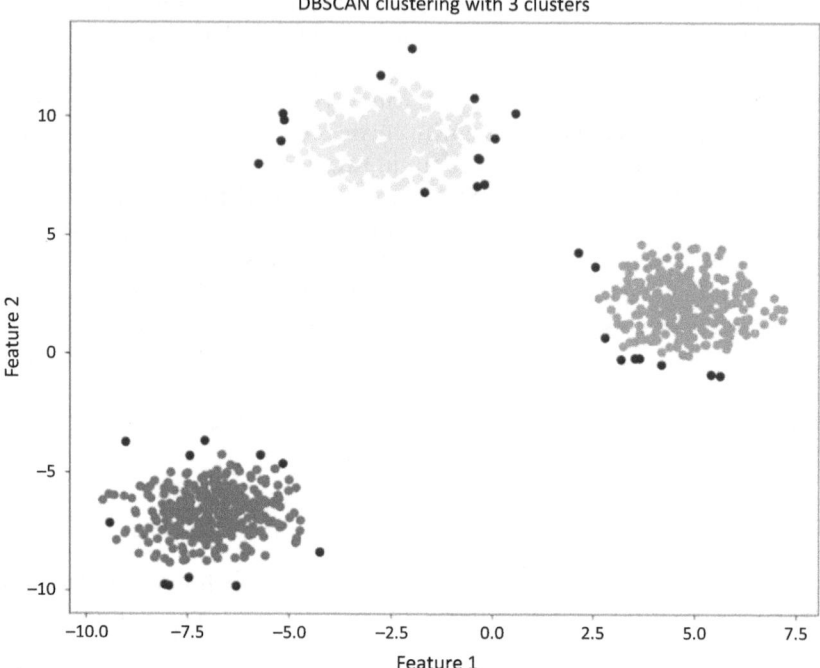

**Fig. 2.5** DBSCAN: plotting the final clusters

occupation, etc. In most cases, we either lack some of these nodes or we have not recorded enough attributes to generate reliable new connections (or groups) among the users. In graph-based ML methods, we use the term feature propagation to refer to a method that may infer or transmit missing node features in our graph. These methods seek to estimate and predict the missing features by making use of the information that is already present in the graph structure and the observed features of the neighbouring nodes (Fig. 2.6).

In mathematical terms, feature propagation can be formulated as a graph inference problem. Let us consider a graph $G = (V, E)$, where $V$ represents the set of nodes and $E$ denotes the set of edges. Each node $v \in V$ has a feature vector $x\_v \in R^\wedge d$ associated with it, where $d$ is the dimensionality of the feature space. Suppose that for some nodes, denoted

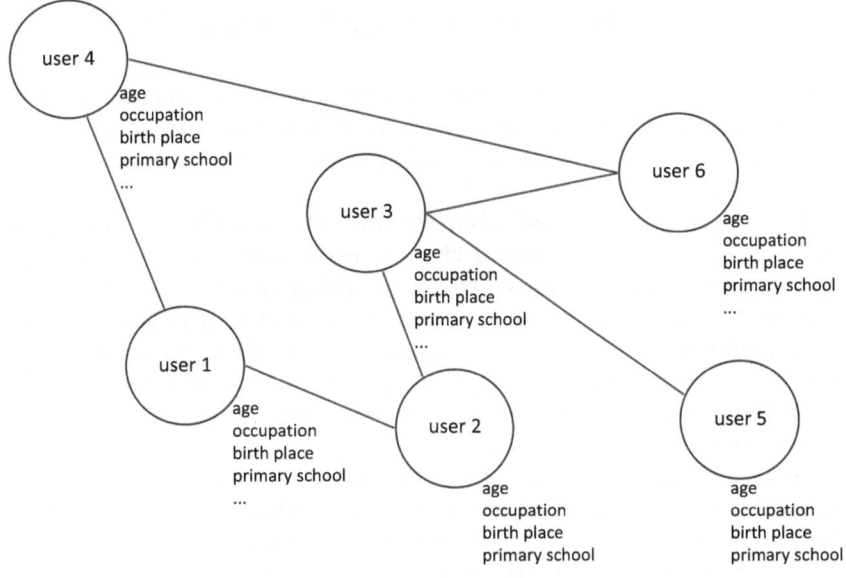

**Fig. 2.6** Rudimentary representation of a network graph depicting relationships on a social media platform

as $V\_m \subseteq V$, the features $x\_v$ are missing or incomplete. The goal is therefore to estimate the missing features for nodes in $V\_m$ based on both the characteristics that have already been observed and the topology of the graph. Feature propagation methods often use a neighbourhood aggregation approach, in which the characteristics of a node are modified by taking into account the characteristics of the nodes that are immediately adjacent to it. In most cases, the process of propagation is carried out iteratively until convergence is achieved or a preset stopping criterion is satisfied.

The **graph convolutional network** (GCN) is widely used for feature propagation. This method aggregates features from neighbouring nodes by first using a linear transformation followed by a non-linear activation function. It then uses the revised features in an iterative process aimed at improving the estimation of missing features. The GCN's approach to propagation can expressed by the following equation:

$$x\_v^{\wedge}(t+1) = \sigma\left(\Sigma\_\{u \in N(v)\}\left(W x\_u^{\wedge}(t)+b\right)\right)$$

Here, $x\_v^{\wedge}(t)$ represents the feature vector of node $v$ at iteration $t$, $N(v)$ denotes the set of neighbours of node $v$, $W$ and $b$ are learnable parameters, and $\sigma$ is an activation function such as the Rectified Linear Unit (ReLU).

Feature propagation methods make efficient use of the network's local structure to estimate the missing characteristics. They do so by repeatedly updating the node features based on the information found in the surrounding neighbourhood. This approach to estimating features is especially useful when analysing social media content. Here, the relationships between users on social media platforms provide essential context from which feature propagation methods can deduce the characteristics and actions of the users themselves.

As an example, let us imagine a simple case in which a social media network consists of three nodes, $A$, $B$, and $C$. While node $A$ and node $C$ have no features, node $B$ has an observed feature vector, $[0.5, 0.3]$. From the structure of the graph, we can deduce that nodes $A$ and $C$ are related to node $B$. During the initial cycle of feature propagation, therefore, we can update nodes $A$ and $C$ with their respective features by aggregating the features of node $B$:

$$x\_A^{\wedge}(1) = \sigma\left(W x\_B^{\wedge}(0)+b\right)$$
$$x\_C^{\wedge}(1) = \sigma\left(W x\_B^{\wedge}(0)+b\right)$$

Subsequent cycles will give us the opportunity to improve our estimations using the most recently observed characteristics of the neighbouring nodes. This procedure is repeated until convergence is achieved, at which point, the features of nodes $A$ and $C$ have been estimated.

To illustrate this process, the following code snippet depicts a simple implementation of feature propagation in Python using GCN with the NetworkX and NumPy libraries:

```
import numpy as np
import networkx as nx
import scipy.sparse as sp
from scipy.sparse.linalg import eigsh

def feature_propagation(graph, observed_features, num_iterations):
    adjacency_matrix = nx.adjacency_matrix(graph)  # Obtain the adjacency
matrix of the graph
    num_nodes = graph.number_of_nodes()
    feature_dim = observed_features.shape[1]

    # Normalize adjacency matrix
    adjacency_matrix = sp.eye(num_nodes) + sp.csr_matrix(adjacency_matrix)
    degree_matrix = np.array(adjacency_matrix.sum(axis=1)).flatten()
    degree_matrix_sqrt = sp.diags(1.0 / np.sqrt(degree_matrix))
    normalized_adjacency =
degree_matrix_sqrt.dot(adjacency_matrix).dot(degree_matrix_sqrt)

    # Initialize feature matrix with observed features
    features = observed_features.copy()

    # Initialize weight matrix and bias term
    weight = np.random.normal(size=(feature_dim, feature_dim))
    bias = np.zeros(feature_dim)

    # Perform feature propagation iterations
    for iteration in range(num_iterations):
        # Compute the updated features using the GCN propagation equation
        updated_features = normalized_adjacency.dot(features).dot(weight) +
bias

        # Apply a non-linear activation function (e.g., ReLU) to the updated
features
        updated_features = np.maximum(updated_features, 0)

        # Update the feature matrix for the next iteration
        features = updated_features

    return features

# Example usage
graph = nx.Graph()
graph.add_edges_from([(0, 1), (0, 2), (1, 2)])
observed_features = np.array([[0.5, 0.3], [0.0, 0.0], [0.0, 0.0]])
num_iterations = 5

estimated_features = feature_propagation(graph, observed_features,
num_iterations)
print(estimated_features)
```

Code snippet 2.6 Feature propagation with the NetworkX and NumPy
Python libraries

This script returns the following data, which estimate the vectors of the missing features within the network (initially set as $[0.0, 0.0]$):

$$\begin{bmatrix} [0. & 0.00145648] \\ [0. & 0.00145648] \\ [0. & 0.00145648] \end{bmatrix}$$

In the example given above, we start by defining a simple graph with three nodes and two edges using the NetworkX library. The *feature_propagation* function uses the graph, the observed features, and the number of iterations as its inputs before initialising the feature matrix, weight matrix, and bias term. It then performs feature propagation iteratively, using the GCN propagation function to update the features of each node based on its neighbours' features. After each iteration, a non-linear activation function (ReLU) is applied to the updated features. This process continues for the number of iterations that have been specified and returns the final estimated features.

### *Analysing Connections*

We can choose from a number of powerful methods if we want to analyse social media data that are organised as a network. Indeed, these methods are particularly powerful when we are studying the connections within the network. A common method used in such analyses, for example, is **node centrality**. This method identifies the most important nodes in a network based on their position in relation to the centre of that network. In this context, three especially useful centrality algorithms are **degree centrality**, **betweenness centrality**, and **eigenvector centrality**.

**Degree centrality** focuses on a node's connections (i.e., number of edges) within a given network. Nodes that are directly connected to a greater number of other nodes have higher degrees of centrality. These nodes are more central to the network because they have greater scope to influence the flow of information and interactions within the network. We can determine the degree centrality of a node $v$ in an undirected network using the following mathematical formula:

$$\text{Degree Centrality}(v) = \frac{(\text{Number of edges connecting to node } v)}{(\text{Total number of nodes} - 1)}$$

We can also deploy two variants of degree centrality when we want to analyse directed networks. While "in-degree centrality" counts incoming edges, "out-degree centrality" counts outgoing edges.

**Betweenness centrality** identifies those nodes that are essential intermediates between distinct portions of the network. A node has a high betweenness centrality if it is located on a large number of the shortest paths that connect it to other nodes. A high level of betweenness centrality enables shows that a node has considerable influence on the flow of both communication and information between various areas of the network. We can determine the betweenness centrality of a node $v$ using the following expression:

$$\text{Betw Centr}(v) = \frac{\Sigma(s \neq v \neq t)\left(\text{Number of shortest paths from } s \text{ to } t \text{ passing through } v\right)}{\left(\text{Total number of shortest paths from } s \text{ to } t\right)}$$

where the variable $s$ represents the source node in the network, $t$ is the target node, and $v$ is an intermediate node (vertex) that is neither $s$ nor $t$. This method counts the number of shortest paths from $s$ to $t$ that pass through $v$. The betweenness centrality of node $v$ is calculated by adding the fraction of shortest paths passing through $v$ over all pairs of nodes $(s, t)$ in the network, excluding $v$ itself.

The importance of a node within its network is determined using **eigenvector centrality**. This method considers both the direct connections of a given node and the centrality of its neighbouring nodes. If a node is connected to other central nodes, then that node is also regarded as central. The primary eigenvector of the network's adjacency matrix is used to determine each node's score. The following equation describes the eigenvector centrality for a node $v$:

$$x_v = \frac{1}{\lambda} + \sum_{t \in M_v} x_t$$

where $M_v$ represents the set of all nodes neighbouring $v$, $t$ is the target node, $x_t$ is the eigenvector centrality of node $t$, and $\lambda$ is a normalisation constant to ensure that the centrality scores are scaled correctly.

Once the centrality and, therefore, the hierarchical position of a certain node has been established, we can turn our attention to analysing the links between nodes. We can, for example, estimate the likelihood of a connection between two nodes in a network by using **link prediction** algorithms. These algorithms study patterns in what connects the other nodes in the network and use a variety of methods to predict further connections. While, on the one hand, predictions can be based on similarities between the nodes (known as similarity-based methods), other predictions are based on the structure of the network (known as network-based methods). And, thanks to these methods, link prediction algorithms enable us to forecast prospective connections or partnerships between social media users.

**Network motif** algorithms search for recurring patterns, such as triangles and other sub-graphs, in the connections that exist within a complex network. We might want to use a network motif algorithm, for example, to identify specific users who frequently engage with the same group of other users, enabling us to posit likely future groupings.

The final class of algorithms I would like to mention in the context of ML techniques for forming and analysing groups are known **community detection** algorithms. We can use these to identify patterns in how a network is connected. Rather than using patterns in users' behaviour or characteristics as a basis for suggesting new groups that users might join, that is, community detection algorithms locate groups or communities that already exist within that network. While, in some cases, these groups may have no formal structure, name, or perceived group identity, they may still already operate as a community of sorts. In this sense, these algorithms are much better at performing analytical rather than predictive tasks. Examples of this class of algorithm include **walktrap community detection**, which finds the densely connected parts of the network (i.e., the communities) by undertaking a random walk, and **surprise community detection**, which evaluates the quality of a network's distribution into communities by using a probabilistic analysis focusing on surprise (measure of disconfirmed expectations with regard to the likely outcome). In this latter approach, nodes are passed from one community to another in order to improve the surprise score.

Before we leave community detection algorithms, however, I shall discuss what are, perhaps, the two most popular of them: the **Louvain** and **Leiden** algorithms (Traag et al. 2019).

The **Louvain** method (Blondel et al. 2008) is based on the concept of modularity (Newman and Girvan 2004) to provide a powerful community detection method that maximises the expected and actual number of edges in a given network. The algorithm starts by partitioning the network, within which each node represents a community. It then passes the nodes from one sub-group to the other, generating different versions of the network (as temporary aggregations of nodes). It repeats this routine until the quality function for modularity is optimised, meaning that it cannot be improved any further.

The Louvain algorithm is one of the most commonly used methods for identifying communities in large networks, as it tends to be fast and perform well. It can, however, generate communities that are poorly connected. Traag et al. (2019) therefore proposed the **Leiden** algorithm as a means to overcome the shortcomings of the Louvain algorithm. The Leiden algorithm is based on the **smart local moving** algorithm (Waltman and Van Eck 2013). Much like the Louvain method, it starts by partitioning the network, but rather than merely creating two subgroups, it repeatedly partitions the network into smaller subgroups. It seeks to refine the quality of connections between communities within the network by generating local connections between subgroups that are within the same community. One partition may therefore be rearranged into a number of sub-groups within the same community. Nodes are then passed from one sub-group to another within the network as a whole until no further improvements are possible.

### *Hierarchical Navigable Small Worlds (HNSW)*

Hierarchical Navigable Small Worlds (HNSW) provides us with yet another powerful method for finding similarities within a network. HNSW was developed relatively recently as a variant of the Navigable Small Worlds (NSW) algorithm, which, in turn, is derived from the concept of the Small World Graph. HNSW works by conducting a vector similarity search using approximate nearest neighbour (ANN) algorithms in high-dimensional spaces. It improves on NSW by adding hierarchical layers to the data structure, thus enabling HNSW to search for nearest neighbours both more quickly and efficiently.

The fundamental driving force behind HNSW is that it generates a network in which every node is considered in a multi-dimensional space. As we see in Fig. 2.7, HNSW spatially arranges nodes across a number of layers. Each group of nodes within a layer constitutes a small world in and if itself, with adjacent nodes linked to one another.

HNSW organises a graph into hierarchical layers, where the lower levels contain a larger number of datapoints and the higher layers generally have

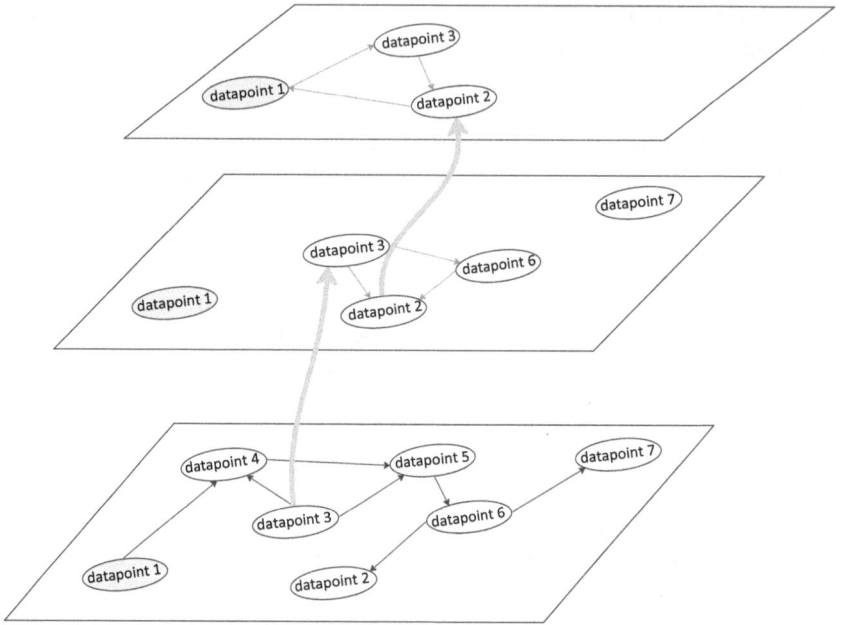

**Fig. 2.7** Layering and traversing in hierarchical navigable small worlds (HNSW)

fewer, all of which are grouped in subsets. The data in each layer are characterised by degrees of refinement (or coarseness). The algorithm seeks out the nearest points, linking them both within and across layers. When it searches between layers for the closest points, therefore, it can do so both quickly and effectively, without needing to compute distant points that are not in the same neighbourhood.

To summarise, the methods and concepts explored in this chapter should help the reader to better contextualise the examples and discussions that follow in the next chapters.

## References

Bhargava, Y. Aditya. 2016. Grokking Algorithms. An illustrated guide for programmers and other curious people. Manning Publication Co. Shelter Island, NY, USA.

Bae, S. (2019) *JavaScript Data Structures and Algorithms: An Introduction to Understanding and Implementing Core Data Structure and Algorithm Fundamentals.* Hamilton, ON: Apress.

Blondel, V.D., Guillaume, J.-L., Lambiotte, R. and Lefebvre, E. (2008) 'Fast unfolding of communities in large networks', *Journal of Statistical Mechanics: Theory and Experiment*, 10, p. P10008.

Bronstein, M. (2022) Feature Propagation is a simple and surprisingly efficient solution for learning on graphs with missing node features. Medium available online: https://towardsdatascience.com/learning-on-graphs-with-missing-features-dd34be61b06 Last accessed: 26 February 2024.

James, J. (2022). 'Data Never Sleeps 10.0', *Domo Blog*, 1 June. Available at: https://www.domo.com/data-never-sleeps (Accessed: 27 February 2024).

Newman, M.E. and Girvan, M. (2004) 'Finding and evaluating community structure in networks', *Physical Review E* 69(2), p. 026113.

Onafuye, R.A. (2021) *The Production of Phygital Social Spaces: A Study on the Influential Factors that Foster New Interactions amongst Second-Generation British Nigerian Youth*. PhD thesis. University of Hertfordshire. Available at: https://uhra.herts.ac.uk/handle/2299/25250 (accessed 30 July 2023).

Traag, V.A., Waltman, L. and Van Eck, N.J. (2019) 'From Louvain to Leiden: Guaranteeing well-connected communities', *Scientific Reports*, 9(1), p. 5233.

Vaswani, A. *et al.* (2017) 'Attention is all you need', *Advances in Neural Information Processing Systems*, 30. Available at: https://papers.nips.cc/paper_files/paper/2017/file/3f5ee243547dee91fbd053c1c4a845aa-Paper.pdf (accessed 27 February 2024).

Waltman, L. and Van Eck, N.J. (2013) 'A smart local moving algorithm for large-scale modularity-based community detection', *The European Physical Journal*, B 86, pp. 1–14.

# CHAPTER 3

# Accidental Collectives

**Abstract** This chapter presents an insight into how collectives are built around the use of online apps and social networks. The chapter explores how large collectives emerge as the accidental result of individuals' quest for self-affirmation and social recognition among a relatively limited group of contacts. The individual user of social media apps shares their daily activities and achievements with a small community of followers in order to receive their appreciation in the form of comments, likes and shares. Whilst the user and their community operate under the presumption that such data are limited and restricted to boundaries of their own network, the software that runs the entire digital system harvests and manipulates their data, aggregating them with other networks to generate trends and improve the system on offer. Such machine learning systems operate as overarching agencies working across communities, contacts and individuals, constructing larger, interrelated and anonymous collectives that are not visible or accessible to the individuals who form them. The software is the only entity with utter control of such collectives.

Through a number of case studies, this chapter explains how such systems work and operate silently in the background of our daily activities. It provides an account of how the private and the public life of individuals are becoming increasingly mediated by software without their full awareness.

S. Carta, *How Computers Create Social Structures*,
https://doi.org/10.1007/978-3-031-62852-8_3

This chapter presents an analysis of some the algorithmic logics that run the transition between private and public lives of individuals and construct large software-driven collectives. This analysis aims at exposing and commenting on some the mechanisms that underpin the silent making of the public life as controlled by machines.

**Keywords**  Urban informatics • Algorithms • Software • Data-driven • Social spaces • Software sorting • Machine learning

## Emergence of Collectives

This chapter is underpinned by the claim that, within a hyper-connected and global society (Vermesan and Friess 2015), collectives emerge as the consequence of the computations operated by software. More specifically, collectives are generated by machine learning-based systems which are underpinned by statistical and mathematical models.

I support this claim through the development of two main points. The first one hinges on the idea that what an individual usually considers their own small social circles is, in reality, characterised by a much wider extent. The first part of this chapter provides a number of examples to illustrate the extent to which personal data are shared with third parties in various ways and with different levels of the individual's awareness. The second part describes how the movement and sharing of data on the Internet are governed by algorithms of which main goals are varied, from the optimisation of data, to automatic generation of semantic from data mining. Through their computational tasks, these algorithms generate new networks which are larger and more connected than those originally thought by the individuals who produced the initial data. The second point of my argument is that these collectives are created accidentally, as the by-product of various computational task. The conclusions discuss the emergence of these new collectives and the extent to which they still reflect human behavioural traits through the mediation of computers.

## Perception and Reality of Individual Networks

There is still a significant discrepancy between the perception of our own audience and who has really access to what we share. Many Facebook users, for example, have recently started to be suspicious about the audience that can be seen in each individual's personal wall. Allegedly, a change in the algorithm used by Facebook in 2019 resulted in some users being able to visualise posts and updates primarily by the same 26 contacts within their network. Although this has been dismissed by Facebook's officials (Facebook 2019), this episode serves as a case in point for this discussion. Some initial studies have begun unveiling the real extent of an individual's online network based on visible traces like retweets (Suh et al. 2010) and overall influence (Bakshy et al. 2011). However, recent works have provided evidence of the extent of the invisible reach of an individual real network. Bernstein and colleagues (Bernstein et al. 2013) compared a direct survey among Facebook users with their relative log data, finding that individual users tend to underestimate their real audience by a factor of four (Bernstein et al. 2013). Moreover, they clarify that *"the actual audience cannot be predicted in any straightforward way by the user from visible cues such as likes, comments, or friend count"* (Bernstein et al. 2013). The authors attribute this mismatch to the fact that social media are characterised by being "socially translucent" as systems which *"provide perceptually-based social cues which afford awareness and accountability"* (Erickson et al. 1999) instead of "socially transparent". This notion clearly describes the real relationship between individuals and their social network. As the consequence of attractive, intuitive and simple interfaces, users are prone to believe to have full control of the data that they share and produce within a particular social medium. Relationships occur in the obliviousness that any direct and indirect contact within the social network is mediated by the network itself, with its filters, suggestions and various other options that customise each user's experience within the site.

In addition to this, the consideration of the mismatch between perceived and real network should also include what is intended and eventually obtained by the user. As demonstrated by (Netter et al. 2013), the user faces a perceived, an intended and an actual disclosure of their privacy/sharing settings (Netter et al. 2013). This is related to what the user actively shares across the social networks and what they leave as a trace in the form of past information which can be accessed by other users (Netter et al. 2013).

Social network users act with clear awareness of their small circle of acquaintances, quite often believing that what they see is the entirety of what happens and, sometimes, having the suspicion that something more complex may be happening in the background. As further elaborated in the following section, the real extent of such networks is vastly larger and more connected than one may expect.

## SOFTWARE, APIS AND MACHINE LEARNING

Generally speaking, social networks include platforms to allow developers and advanced users to retrieve certain data from the social media connecting them to other websites and applications. Such platforms usually take the form of Application Programming Interface (API) and have a key role in integrating and connecting social data, feeds and users through the Internet. APIs are a central technology to understand the relationship between online users, their social behaviour and the formation of collectiveness. A comprehensive overview of the technological framework which underpins APIs and, more in general the tools widely available to analyse data in social media, is provided by (Batrinca and Treleaven 2015). In particular, the two authors refer to sentiment analysis (Pang and Lee 2008) as an approach which is: "*about mining attitudes, emotions, feelings—it is subjective impressions rather than facts*" (Batrinca and Treleaven 2015). Batrinca and Treleaven break down sentiment analysis into sub-aspects including context, level, subjectivity, orientation and strength (Batrinca and Treleaven 2015) and relate it to supervised learning algorithms like Naive Bayes (NB), Maximum entropy (ME), Logistic regression (LR), Latent semantic analysis and Bag-of-words (Batrinca and Treleaven 2015). In this case, statistical models are used to infer the emotions and feelings of users and produce analyses and predictions of future possible behaviour. Such an approach is a strong feature of the interaction machine learning-human behaviour, whether it is applied to the finding of personal data from online activities or automatic classification of users. The following examples illustrate some the mechanism by which machine learning algorithms are deployed to compute human traits in datasets.

### *Inferring Personal Data*

Even with encrypted networks, it is possible to infer personal data through the use of social apps. Atkinson, Mitchell, Rio and Matich, for example, demonstrated that users' activity can be inferred by analysing information extracted from encrypted WiFi networks (Atkinson et al. 2018). By processing the limited data available from an encrypted network (inter-arrival time between frames, the direction from broadcast sender to the receiver and the frame size) in the most popular mobile apps, this study illustrates how information about lifestyle, religion or interests and hobbies can be deducted (Atkinson et al. 2018:552). For example, Grindr appears to be one the apps that allows more information to be extracted, including gender, marital status, income, and information about dating habits and sexuality, as well as information about the own mobile device like its capability (Atkinson et al. 2018:550). It is important to note that this study shows that it is possible to find all this information from individuals simply looking at side-channel data and data available to anyone with the appropriate technology and expertise. There is no attempt to directly break the encryption of access data directly (Atkinson et al. 2018:552).

A slightly more intrusive method is shown by (Wang et al. 2015), where packet-level traffic is analysed using data leaks in the side-channel to extract information about specific user's patterns in data transfer outside the encryption (Wang et al. 2015:433). An attacker could easily sniff data to deduct texts, pictures and streaming from browsers or messages, contacts, posts and pictures from social networks like Twitter or Facebook (Wang et al. 2015:436).

Attackers can even recognise which app in one's mobile device uses fingerprint recognition technologies, inferring which app are most data-sensible and therefore more potentially interesting from the attacker point of view. Taylor and colleagues (Taylor et al. 2016) developed AppScanner: an app that generates an automatic fingerprinting of apps within an encrypted Network. By using a combination of Support-vector classifier (SVC) and Random Forest (RF) algorithms, this framework is able to return an accurate estimate of which app a particular individual is using by analysing their network traffic.

On a higher level of traffic analysis, behavioural traits can be inferred by simply looking at the IP headings. To demonstrate this, (Saltaformaggio et al. 2016) developed NetScope: a software that analyses data traffic within a network, for example transfer rates, packet exchanges, and data transfer to identify a specific app. This project is underpinned by the idea that the data traffic varies significantly depending on the on-line behaviour of the user (scrolling through the news feed as opposed to posting new images on social network). NetScope works with a K-means clustering algorithm (see Chap. 2) that allows to divide the measurements of the behaviours in the network traffic into clusters based on their position and proximity to the clusters' centres (Saltaformaggio et al. 2016:4). By means of clustering, this algorithm returns as output a distinctive set of values for each cluster, which then become easily classifiable (Saltaformaggio et al. 2016).

## Automatic Classification of People

Through their on-line behaviour, people are classified automatically by algorithms. Within a digital environment, people's actions are considered through the data that represent them. Data are then collected, organised and classified in order to generate a meaningful sense of them. Interestingly, human actions are turned into data that are computed by means of machine learning tools, then translated back into representation of human behaviour. It is important at this point of the argument to focus on the key role that classifiers play in generating categories and sorting the observed behaviours into these categories. An illustrative example is the automated establishment of individuals' political orientation and ethnicity through the observation of social behaviour on social media. For example, (Pennacchiotti and Popescu 2011) demonstrated how this is possible by deploying a machine learning technique to analyse contents and structure of Twitter feeds. By using a Gradient Boosted Decision Trees—GBDT (Friedman 2001), Pennacchiotti and Popescu demonstrated how four types of information can be extracted: profile features, messaging behaviour, the linguistic content of messages and social network information (Pennacchiotti and Popescu 2011:282). In their experiment, they illustrated how the algorithm classified a sample of more than 10,000 users into four groups that include: democrats, republicans, African-Americans and Starbucks fans (Pennacchiotti and Popescu 2011:284).

A more explicit example is perhaps when personal traits can be accurately predicted by simply looking at the snapshot of one's personal device. In particular, (Seneviratne et al. 2014) demonstrate how supervised algorithms (see Chap. 2) can be used to infer personal information, including religion, parental and relational status, spoken languages and overall preferences and interests. By using binary (i.e. yes/no) Support Vector Machine (SVM) algorithms, the app (Apptronomy) Seneviratne and colleagues built was able to analyse data from about 200 mobile users and infer with an accuracy of 85% the likelihood of features including country of origin/residence, language, partnership (single/not-single), religion (Christianity, Islam, Hinduism, Buddhism) and parenthood (parent, not parent) (Seneviratne et al. 2014:3).

Some algorithms are also able to discern real people from bots. Here again, the method that can be used to operate such decision is ML-based. Through various versions of Twitter APIs, Wang collected datasets for three weeks with more than 25,000 users with approximately 500,000 tweets and a network of around 49,000,000 of inter-related users (followers/friends) (Wang 2010:340). By using three graph-based features to plot a graph of the mutual relationships among users and three content-based features to extract contents from the tweets (see Chap. 4 for more details), Wang (2010:337) was able to determine through a Bayesian classifier which tweets were generated by bots from the genuine ones with a good degree of accuracy.

Furthering this idea, (Lee et al. 2010) demonstrated how honeypots (virtual and safe environments where various threats can be tested simulating the work of possible malicious attackers) can be used in social sites like Twitter or MySpace to identify spammers and malicious users within social networks. Within the context of their work, social honeypots are defined as: *"information system resources that monitor spammers' behaviors and log their information (e.g., their profiles and other content created by them in social networking communities)"* (Lee et al. 2010:436). Lee and colleagues developed an interesting work that illustrates how classifiers can be used to discover spam profiles on Twitter when the algorithm is deployed over a large number of unknown (random, real) profiles.

Such approaches have been used to automatically detect negative behaviours such as cyberbullying. A good example is the work of (Galán-García et al. 2016), where the authors have found a way to associate a fake profile to a real one within the same social network. Their work is underpinned by the hypothesis that behind each trolling account there is a real one related to the fake account. The real account allows the troll to stay updated about the social activities around the fake account and act responsively. In this work too, classifiers including Random Forest (RF), K-Nearest Neighbor (KNN) and Sequential Minimal Optimization (SMO) were used to test the most efficient and reliable method to evaluate the authorship of the tweets and assign them to the real/troll user. The algorithms were trained with around 1,900 tweets from about 20 Twitter accounts, and the method, ultimately applied to a real-case scenario, resulted in the recognition of a troll over more than 17, tweets analysed (with 43 tweets identified as from trolling profile) (Galán-García et al. 2016:49).

## Prediction of Behaviour and Grouping

Machine learning algorithms can be used to predict individual's behaviour in social networks. For example, in the work of (Benchettara et al. 2010), the prediction is considered as a *"two-class discrimination problem"*, which thus allows to *"apply classical supervised machine learning approach for learning prediction models"* (Benchettara et al. 2010:327). The prediction is related to three criteria: Dyadic (which evaluates a value for each link for each couple of vertices in a graph) Vs. Structural approaches (which evaluates rules of evolution of sub-graphs in a network); Topological (related to characteristics like degree, common neighbours or distance between nodes) Vs. Node features based approaches (related to the characteristics of the node itself); and Temporal vs. non temporal approaches (relating to the time of link formation in a network) (Benchettara et al. 2010:327). The prediction model used by Benchettara and colleagues is based on the discrimination of linked classes from unlinked ones to determine the likelihood of a certain link to be formed within a network.

Machine learning can be used to predict the likelihood of one individual to be interested in partaking a specific group of other likeminded users or other individuals with shared interests. A project developed by (Amershi et al. 2012) illustrates how their app "ReGroup" uses end-user interactive machine learning to generate collectives within social media in general and Facebook in particular (Amershi et al. 2012:21). ReGroup uses a Naïve Bayes classifier whereby the probability of combining two users is computed by a Bayesian equation with features (gender, city, workplace etc.), memberships of other groups and probabilities of belonging to groups are the parameters. The algorithm computes the probability of acceptance of the group and social "friendship" outputting a certain value. On the basis of this value, groups are suggested and classifier-driven collectives may emerge as a result.

More specifically, algorithms can be deployed to ascertain the influence of a certain node in a social network (this being a user or a topic of discussion for example) over the other members of that group. This idea is quite explicit in the work of (Tang et al. 2009) and defined as "social influence analysis" (Tang et al. 2009:808), where they have been able to identify the key nodes within a network (that is those with more influence over the others) and the reciprocal influence of neighbouring nodes. To do this, Tang and colleagues proposed the Topical Factor Graph (TFG) model, designed to: "incorporate all the information into a unified probabilistic model" (Tang et al. 2009:809). Among other parameters, this model includes factors like the number of nodes in a social network, number of edges, number of topics, hidden vectors of representatives for all topics on nodes, the probability for one topic to be generated by a specific node, and the social influence of a node onto another one. To test the TFG model, Tang et al. deployed their algorithm on a database of journal articles authors and their relative citations generating a network of new links and relationships based on mutual social influence (Tang et al. 2009:813–814).

To conclude, projects like those mentioned here suggest that new collectives are generated almost accidentally by algorithms. The main goal of algorithms is to mine data and generate meaning out of them. By doing this, they generate larger and global collectives that are software-driven, albeit of secondary importance.

## Happn

This section illustrates how a popular social app works with data aggregations and the accidental formation of networks. Happn represents an interesting case for it allows users to combine the date-stamp of their daily activities with a specific path taken. The inclusion of location data into dating apps opens up new venues of investigation and potentials in urban studies (Ma et al. 2017). In particular, scholars have recently focused on leveraging spatial data based on online and offline activities of users (La Morgia et al. 2018). The links between users that the algorithms of this app generate are broadly unknown to users as this type of information is automatically generated by computers in a machine-readable format. Such data can be accessed by people (e.g. data brokers, data scientists, hackers etc.) yet, this would require a certain level of computing knowledge and some algorithmic techniques. The work of (Di Luzio et al. 2018) is very useful in demonstrating the extent to which dating apps like Happn generate a network of users which is much larger than what the individual members can see. As expected, the app has a number of security mechanisms that filter and protect personal information about each user involved. For example, the app indicates which users have crossed paths in a particular time. This information is approximated on purpose to protect the privacy of the users. The real distance is turned into categories (e.g., within 300 m, within 500 m etc.), yet Di Luzio and colleagues showed how an attacker may be able to retrieve more accurate details about the exact position of users (Di Luzio et al. 2018:840). Moreover, they also demonstrate how to de-anonymise users and their distance by means of *trilateration*, by which a polygon is created using different measurement points as vertices and ensuring that the target point (the exact position of the targeted user) is contained in it (Di Luzio et al. 2018:841). With this technique, they demonstrated how the initial cross-path distance indicated by the app can be re-measured with more accuracy: "*The initial trilateration [...] [was] 79:66m. The succeeding iterations progressively increase the accuracy of the result, pinpointing the target to an area roughly the size of a basketball court with an error of 6:21m*" (Di Luzio et al. 2018:841).

This experiment becomes significant if we consider the implications of the potential security breach of personal data. De Luzio and colleagues tracked and profiled over 10,000 users, uncovering "*their daily routines, their home and work addresses, where they like to hangout on weekends, and who their friends, their relatives, and their colleagues are. Through Happn we not only expose their private lives, but their social connections as well*" (Di Luzio et al. 2018:841).

Flipping the perspective of this findings from the user's to the software viewpoint, we can appreciate the extent to which the app has centralised and granular data about each user and can, therefore, be used to infer new data about a particular individual, as well as groups of people. These data-based collectives can be filtered by any of the attributes that users specify in their profile, including gender, religion, interests, whereabouts, place of work and residence etc. In principle, the app is the only entity able to oversee all the data produced by users and to form new groups.

## New Collectives

Some of the online systems observed in the previous sections, including Grindr, AppScanner, NetScope and ReGroup generate extra layers of inferred information about individuals and their network. By creating new associations, links and connections between data and users, they create entire new networks which are invisible to the individual users to which they partially pertain, and from which they originate. Users may only see some glimpse of these new networks through their effects. For example, in the form of ad-hoc adverts or suggestions of products to buy, holidays to book, or new friendships.

There two main considerations that result from this study. The first one is a possible description of the configuration of the mutual relationship

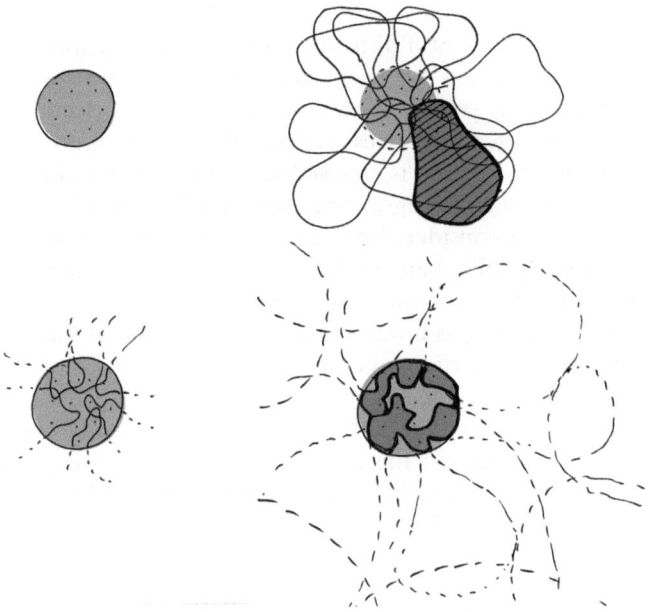

**Fig. 3.1** Conceptual production of the configuration of a personal network

between an individual's personal network and its real extent. Figure 3.1 represents the inner circle (dark grey) which is known to the user and contains their own acquaintances (light grey spots). This included and well-defined shape is intersected at various points by a second network of which real shape and extent are not clearly defined (dashed continuous line). The exact shape and size of the larger network are difficult to describe as they depend on a number of variables within which are part of complex data analytics and ML-driven systems.

Users are aware of the presence of larger networks created by algorithms and somehow of their action and influence in their everyday use of social media. People seem to perceive the ordinary effects of the work these clustering, filtering and sorting algorithms as documented by Taina Bucher (Bucher 2019). Users react to the intrusion of the larger network in their own small circle in a variety of ways, from declaring their "*frustration of not getting any 'likes' to the strange sensation of thinking 'who are these people who suddenly show up'?*" (Bucher 2019:42). People's reaction to the interference of larger network allows their personal network to temporarily expand to create a new form of collectiveness which are amorphous, and of which extent if difficult to grasp and to quantify.

Lastly, the intersection between larger collectives into personal networks shows how the private sphere of individuals transcends into their public life. When larger collectives are created, what the individual considered as personal information is exposed to an unknown and uncontrolled new audience for which an algorithm established a possibly valuable connection. What it was considered private, or shared with a small number of contacts becomes public, namely shared with other individuals of whom there is no information or control (Fig. 3.2).

The second outcome is a reflection on the accidental nature of these new collectives that emerge as a consequence of the work of algorithms. From a social point of view, larger collectives can be regarded as an accidental occurrence as they are not the software's primarily objective, yet they are formed. As seen in the case of ReGroup, for example, a Naïve-Bayes classifier is used to evaluate a probability of friendship between two individuals. The interested parties will be able to see the results of this

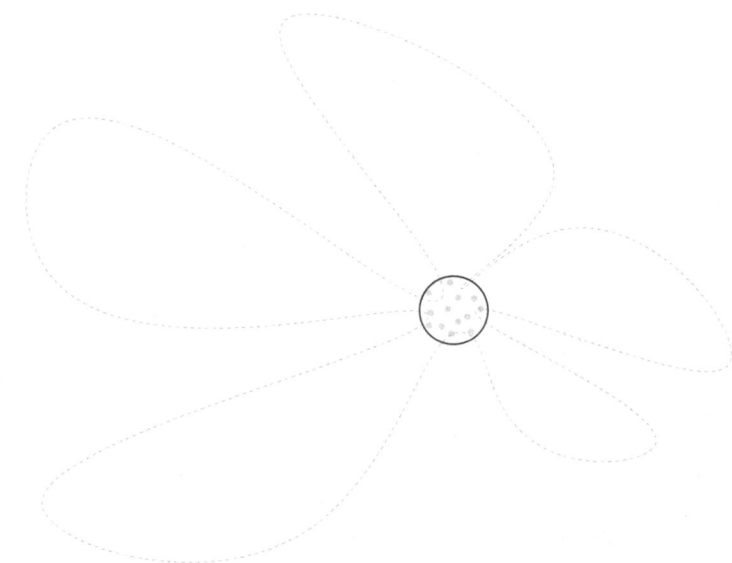

**Fig. 3.2** Configuration of personal network (illustrated with the thick line) and its real extent (represented with dash lines)

computation through the suggestion of a connection. They will only experience a small portion of the new collective created by the algorithm. The entire extent of the new collective is never manifested as it assumes the form of computational outputs scattered in a number of databases to be retrieved when and if needed. The algorithm calculates a percentage of affinity between two users. These values are used to suggest new links between users. After a number of iterations, individuals will start to connect, expanding their own networks. However, if considered altogether, the totality of user profiles and their respective affinity values form a new collective which is invisible in its entirety to everyone but the software itself (Fig. 3.3).

The software's task is to compute inputs and return outputs, suggesting actions to users. In this sense, the software has no intention of producing or even considering the collectives that its computations generate as a side effect. These new collectives are created by machines that have no interest in them, yet they are invisible and inaccessible to humans who are the main reason for their existence.

In this sense, these vast collectives lay on a separate dimension from the one of the users that compose them. At times, these new collectives may interfere with the personal network of the users to deliver suggestions, adverts or predictions, yet never fully integrating with the inner networks of each user.

As further elaborated in Chap. 5, this study suggests that more attention should be paid to the nature of the machine behaviour. Software, algorithms, intelligent agents, Machine Learning (ML) and Artificial Intelligence (AI) systems are having an increasing impact on our social experience and

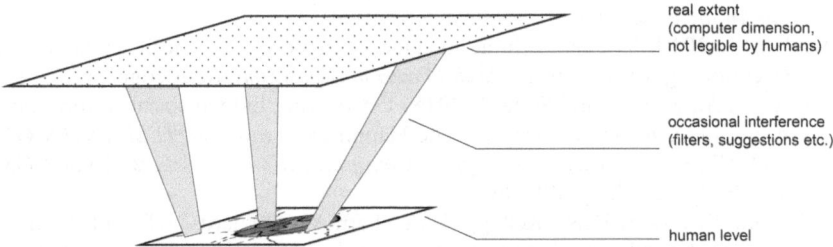

**Fig. 3.3** Conceptual representation of the own personal network. The diagram shows the three levels of network perception. Human level at the bottom, occasional interference in the middle and the real extent (always invisible) at the top

in both our private and public spheres. As pointed out by Iyad Rahwan and colleagues: "*understanding the behaviour of artificial intelligence systems is essential to our ability to control their actions, reap their benefits and minimize their harms*" (Rahwan et al. 2019:477). The algorithmic logics that underpin the creation of these new large collectives should be studied more widely, considered by interdisciplinary teams of architects, designers, historians, and scholars in the humanities and social sciences. In summary, these new collectives should be also understood by other than those who created them in the first place (Rahwan et al. 2019:477).

## REFERENCES

Amershi, S., Fogarty, J. and Weld, D. 2012. Regroup: Interactive machine learning for on-demand group creation in social networks. In: *Proceedings of the SIGCHI conference on human factors in computing systems.* pp. 21–30.

Atkinson, J.S., Mitchell, J.E., Rio, M. and Matich, G. 2018. Your WiFi is leaking: What do your mobile apps gossip about you? *Future Generation Computer Systems* 80, pp. 546–557.

Bakshy, E., Hofman, J.M., Mason, W.A. and Watts, D.J. 2011. Everyone's an influencer: quantifying influence on twitter. In: *Proceedings of the fourth ACM international conference on Web search and data mining.* pp. 65–74.

Batrinca, B. and Treleaven, P.C. 2015. Social media analytics: a survey of techniques, tools and platforms. *Ai & Society* 30, pp. 89–116.

Benchettara, N., Kanawati, R. and Rouveirol, C. 2010. Supervised machine learning applied to link prediction in bipartite social networks. In: *2010 international conference on advances in social networks analysis and mining.* IEEE, pp. 326–330.

Bernstein, M.S., Bakshy, E., Burke, M. and Karrer, B. 2013. Quantifying the invisible audience in social networks. In: *Proceedings of the SIGCHI conference on human factors in computing systems.* pp. 21–30.

Bucher, T. 2019. The algorithmic imaginary: Exploring the ordinary affects of Facebook algorithms. In: *The Social Power of Algorithms.* Routledge, pp. 30–44.

Di Luzio, A., Mei, A. and Stefa, J. 2018. Uncovering hidden social relationships through location-based services: The Happn case study. In: *IEEE INFOCOM 2018-IEEE Conference on Computer Communications Workshops (INFOCOM WKSHPS).* IEEE, pp. 802–807.

Erickson, T., Smith, D.N., Kellogg, W.A., Laff, M., Richards, J.T. and Bradner, E. 1999. Socially translucent systems: social proxies, persistent conversation, and the design of "babble". In: *Proceedings of the SIGCHI conference on Human Factors in Computing Systems.* pp. 72–79.

Facebook. 2019. *No, Your News Feed Is Not Limited to Posts From 26 Friends.* Available at: https://about.fb.com/news/2019/02/inside-feed-facebook-26-friends-algorithm-myth/.

Friedman, J.H. 2001. Greedy function approximation: a gradient boosting machine. *Annals of statistics*, pp. 1189–1232.

Galán-García, P., Puerta, J.G. de la, Gómez, C.L., Santos, I. and Bringas, P.G. 2016. Supervised machine learning for the detection of troll profiles in twitter social network: Application to a real case of cyberbullying. *Logic Journal of IGPL* 24(1), pp. 42–53.

La Morgia, M., Mei, A., Raponi, S. and Stefa, J. 2018. Time-zone geolocation of crowds in the dark web. In: *2018 IEEE 38th International Conference on Distributed Computing Systems (ICDCS)*. IEEE, pp. 445–455.

Lee, K., Caverlee, J. and Webb, S. 2010. Uncovering social spammers: social honeypots+ machine learning. In: *Proceedings of the 33rd international ACM SIGIR conference on Research and development in information retrieval*. pp. 435–442.

Ma, X., Sun, E. and Naaman, M. 2017. What happens in happn: The warranting powers of location history in online dating. In: *Proceedings of the 2017 ACM Conference on Computer Supported Cooperative Work and Social Computing*. pp. 41–50.

Netter, M., Riesner, M., Weber, M. and Pernul, G. 2013. Privacy Settings in Online Social Networks-Preferences, Perception, and Reality (Best Paper Award Nominee).

Pang, B. and Lee, L. 2008. Opinion mining and sentiment analysis. *Foundations and Trends® in information retrieval* 2(1–2), pp. 1–135.

Pennacchiotti, M. and Popescu, A.-M. 2011. A machine learning approach to twitter user classification. In: *Proceedings of the international AAAI conference on web and social media*. pp. 281–288.

Rahwan, I. et al. 2019. Machine behaviour. *Nature* 568(7753), pp. 477–486.

Saltaformaggio, B. et al. 2016. Eavesdropping on {Fine-Grained} user activities within smartphone apps over encrypted network traffic. In: *10th USENIX Workshop on Offensive Technologies (WOOT 16)*.

Seneviratne, S., Seneviratne, A., Mohapatra, P. and Mahanti, A. 2014. Predicting user traits from a snapshot of apps installed on a smartphone. *ACM SIGMOBILE Mobile Computing and Communications Review* 18(2), pp. 1–8.

Suh, B., Hong, L., Pirolli, P. and Chi, E.H. 2010. Want to be retweeted? large scale analytics on factors impacting retweet in twitter network. In: *2010 IEEE second international conference on social computing*. IEEE, pp. 177–184.

Tang, J., Sun, J., Wang, C. and Yang, Z. 2009. Social influence analysis in large-scale networks. In: *Proceedings of the 15th ACM SIGKDD international conference on Knowledge discovery and data mining*. pp. 807–816.

Taylor, V.F., Spolaor, R., Conti, M. and Martinovic, I. 2016. Appscanner: Automatic fingerprinting of smartphone apps from encrypted network traffic. In: *2016 IEEE European Symposium on Security and Privacy (EuroS&P)*. IEEE, pp. 439–454.

Vermesan, O. and Friess, P. 2015. *Building the Hyperconnected Society-Internet of Things Research and Innovation Value Chains, Ecosystems and Markets*. Taylor & Francis.

Wang, A.H. 2010. Detecting spam bots in online social networking sites: a machine learning approach. In: *IFIP Annual Conference on Data and Applications Security and Privacy*. Springer, pp. 335–342.

Wang, Q., Yahyavi, A., Kemme, B. and He, W. 2015. I know what you did on your smartphone: Inferring app usage over encrypted data traffic. In: *2015 IEEE conference on communications and network security (CNS)*. IEEE, pp. 433–441.

# CHAPTER 4

# Automated Spatial Configurations

**Abstract** This chapter illustrates the extent to which software influence the generation of public spaces. In particular, I discuss a number of projects where machine learning techniques and AI-driven methods in general are used to automatically generate buildings, from houses to care home facilities. These examples are used to emphasise the mechanisms through which designers and programmers generate software that eventually determine the ways in which people use physical spaces. Through three projects, this chapter discuss how the elaboration of a set of a-priori design criteria (or objectives) results in the automatic generation of public spaces, common areas and social places where people meet, interact and conduct their public life.

**Keywords** Spatial configurations • Self-organising spaces • Social spaces • Physical spaces • Automation • AI

## FROM CODE TO SPACE

The process of turning programming code into spatial configurations in buildings typically involves the use of computational design techniques. These techniques require the use of algorithms and pieces of software to generate and manipulate digital models of architectural spaces. The

S. Carta, *How Computers Create Social Structures*,
https://doi.org/10.1007/978-3-031-62852-8_4

process typically starts with the definition of design goals and objectives, which are translated into a set of computational instructions and rules. These instructions are then used to generate a range of possible spatial configurations, which are evaluated against a set of design criteria and constraints.

To translate the digital models generated through this process into physical buildings, various digital fabrication techniques can be employed. These methods essentially centre on the use of computer-controlled machinery to convert digital models into real, observable objects. This covers a wide range of techniques, such as CNC milling, laser cutting, and 3D printing. Overall, the process of turning programming code into spatial configurations in buildings involves the integration of computational design and digital fabrication techniques, with the aim of producing buildings that are optimised for specific design goals and constraints. A very simple example of how spatial organisations can be generated using code is included here (Fig. 4.1):

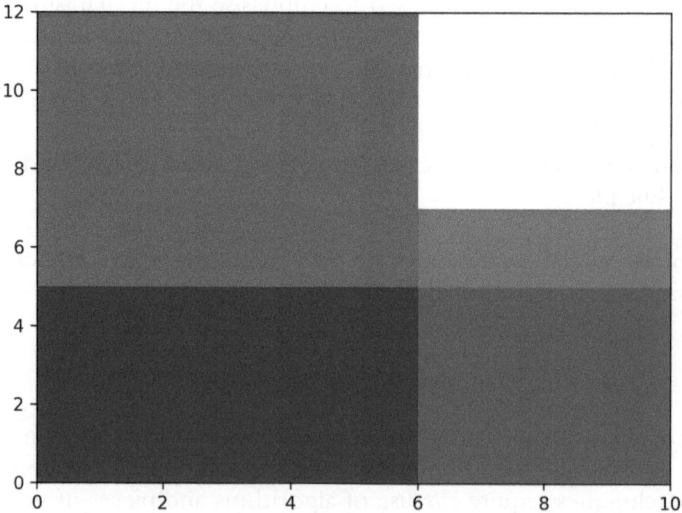

**Fig. 4.1** Example of spatial configuration (floor plan) generated with a simple python code

```
import matplotlib.pyplot as plt
# Define the dimensions of the house
width = 10
length = 12

# Define the different rooms of the house
kitchen = (0, 0, 6, 5)
living_room = (6, 0, 6, 5)
bedroom = (0, 5, 6, 7)
bathroom = (6, 5, 6, 2)

# Create a new figure and axis to plot on
fig, ax = plt.subplots()

# Draw the rooms as rectangles on the plot
ax.add_patch(plt.Rectangle((kitchen[0], kitchen[1]), kitchen[2], kitchen[3],
facecolor='blue'))
ax.add_patch(plt.Rectangle((living_room[0], living_room[1]), living_room[2],
living_room[3], facecolor='green'))
ax.add_patch(plt.Rectangle((bedroom[0], bedroom[1]), bedroom[2], bedroom[3],
facecolor='red'))
ax.add_patch(plt.Rectangle((bathroom[0], bathroom[1]), bathroom[2],
bathroom[3], facecolor='gray'))

# Set the limits of the plot to match the dimensions of the house
ax.set_xlim([0, width])
ax.set_ylim([0, length])

# Display the plot
plt.show()
```

Code snippet 4.1 A very basic spatial organisation

In this example (Fig. 4.1), each room is represented as a tuple with four values: (x-coordinate, y-coordinate, width, height), so that the room kitchen is defined by four values: kitchen = $(0, 0, 6, 5)$. When drawing the individual rooms, we use the 'add_patch' function of the axis to return a patch (area) within the given axes (x, y). In short, this function draws a rectangle per each room following the dimensions given when we defined the rooms. The final step is to shape the rooms to fit the overall size of the house, defined at the beginning with a specific width and length. This ensures that the plot only shows the area within the house envelope. This very simple idea is further elaborated with the concept of self-organising spaces, where code is used establish rules that generates spatial configurations.

## SELF-ORGANISING SPATIAL CONFIGURATIONS

The notion of Self-Organising Floor Plans (SOFP) could be generally defined as the methods employed by designers to generate spatial layouts automatically. These techniques involve the use of machine learning (ML) and optimisation techniques, neural networks (NN), and evolutionary algorithms. The models usually include topological approaches such as Kohonen's self-organizing feature maps (SOMs) and object recognition with deep neural networks. I have explored this notion in detail in (Carta 2021). The concept of self-organisation varies depending on its context. In the study of complex systems, self-organisation is linked to the idea of emergence, where a global order is achieved through interactions among the parts governed by local rules. Key literature on spatial emergence includes the seminal work carried out in the 1990s and 2000s by John H. Holland (1996, 2000, 2002) and Steven Johnson (2002), where, oversimplifying, adaptations to external conditions to a given system drives the emergence of new (spatial) patterns. While Holland and Johnson's work (and that which followed) has been key for us to understand how spatial conditions can emerge primarily based on observation of nature, technological progress specifically in the field of artificial intelligence expanded our understanding of how space can be generated through sophisticated mathematical steps. In fact, we have a long list of important work on self-organisation that includes a large range of methods, from topological approaches to neural networks and genetic algorithms. Among many others, it is important to mention the work of scholars such as (March and Steadman 1971; Shaviv 1987), as well as more recent studies including (Guo and Li 2017; Martin 2005; Merrell et al. 2010; Michalek et al. 2002; Zheng and Yue 2020). Quite established areas include Automated Facilities Layout (Chen et al. 2021; Levary and Kalchik 1985; Liggett 2000; Seehof et al. 1966), Spatial Synthesis (Eastman 1975; Jo and Gero 1998; Veloso and Krishnamurti 2021), Space Planning (Anderson et al. 2018; Brookes and Kaplan 1972), Layout Synthesis (Liggett 2000; Wu et al. 2019) and, even more recently, detailed studies of floor plans composition (Park et al. 2023; Shekhawat 2023).

## Case Study 1: AISLA

With the project Analyse and Improve Spatial LAyout of care homes (AISLA), we developed a novel method to analyse spatial configurations and qualities of existing care homes in the UK with the aim of helping care home managers and those involved in maintaining and regulating health settings to improve building standards. We gathered and combined data derived from the following three sectors: the design criteria of therapeutic environments, net zero strategies and the feedback of care homes users (residents, their families and staff members). We combined such factors into intervention criteria (or actions) to support managers to identify how to improve their buildings. With our tool, they are now able to assess existing care homes and visualise possible options for retrofitting based on net-zero strategies. As one of the outcomes of this work, we developed a web-based app that analyses existing layouts of care homes, providing a number of performance ratings and suggesting points for improvements. For this study, we used the following design criteria:

*Design Criteria for Care Homes*
- Privacy
- Easy navigation & orientation
- Size of building environment
- Size and layout of dining areas
- Provision of specially equipped environment support
- Natural daylight
- Garden & experiencing nature
- Outside views
- Views of nature, biophilic scenes and other positive distractions
- Access to outside space
- Spaces for stimulating activities—Access to Social support—Social interactions
- Presence of art
- Control and choices
- Support of functional abilities—Resident mobility issues
- Comfort—Materials
- Contact with staff
- Air quality
- Safety and security
- Provision of easy access from outside

- Energy efficiency
- Carbon compliance level
- Allowable solutions
- Cost

According to the type of each option and the area of the built environment that each one affects, the design criteria can be further clustered into four main actions:

- Building
- Rooms
- Common areas
- Outdoor & windows

Mutual interactions could occur among criteria of different actions; future research could further investigate the interrelationship amongst all design criteria and the possible impact each criterion has on each other. The online app we developed can be access at: https://aisla.vercel.app/

We developed a Single Page interactive JavaScript application and deployed it on the Web. The app is responsive as it has been designed to work and adapt itself to small screen devices such as smartphones, and to big desktop screens. The app is a step-by-step wizard-like survey that guides the users in answering different sets of questions about different care home features (clustered in Building, Rooms, Communal, and Outdoor), and finally displays the final ranking result. The app breaks down all scores from each factor (e.g. in Fig. 4.2), giving recommendations on possible improvements. The app provides suggestions on how to improve the overall score. Finally, the app allows users to export and save a report summarising scores and improvements. The Technology Stack used is: Next.js (production framework for React apps) https://nextjs.org, Material-UI (User Interface Elements based on the Material Design System by Google) https://mui.com, Recharts (Charting Library) https://recharts.org

We developed the app in a modular way allowing four types of questions: Yes/No, Multiple Choice, Slider, and Number Input. Each type gives the opportunity to specify its own constraints, such as upper and lower limits. The User Interface (UI) elements are standardised to give a consistent look and feel to the app (Fig. 4.3).

# COMMON AREAS

| | | Questions | Answers |
|---|---|---|---|
| **1** | 49 | Questions:<br>Does the care home have small dining rooms of domestic scale that promote social interaction and conversation through the creation of a homely atmosphere? | Answer:<br>yes/no [multiple choice button] |
| | 50 | Questions:<br>How many square meters is the dining area(s)? | Answer:<br>XXX sqm [float] [text box] |
| | 51 | Questions:<br>Is there sufficient space in dining rooms for staff to help? | Answer:<br>yes/no [multiple choice button] |
| | 52 | Questions:<br>Do residents have access to drinks and snacks during the day? | Answer:<br>yes/no [multiple choice button] |
| **2** | 53 | Questions:<br>Does the care home have an adequate number of suitable rooms/spaces for stimulating activities to take place (such as physical activity, music, art & crafts, etc)? | Answer:<br>yes/no [multiple choice button] |
| | 54 | Questions:<br>Does the kitchen(s) layout allow participation in food preparation and social interaction among residents and/or their visitors? | Answer:<br>yes/no [multiple choice button] |

**Fig. 4.2** Example of design criteria clustering

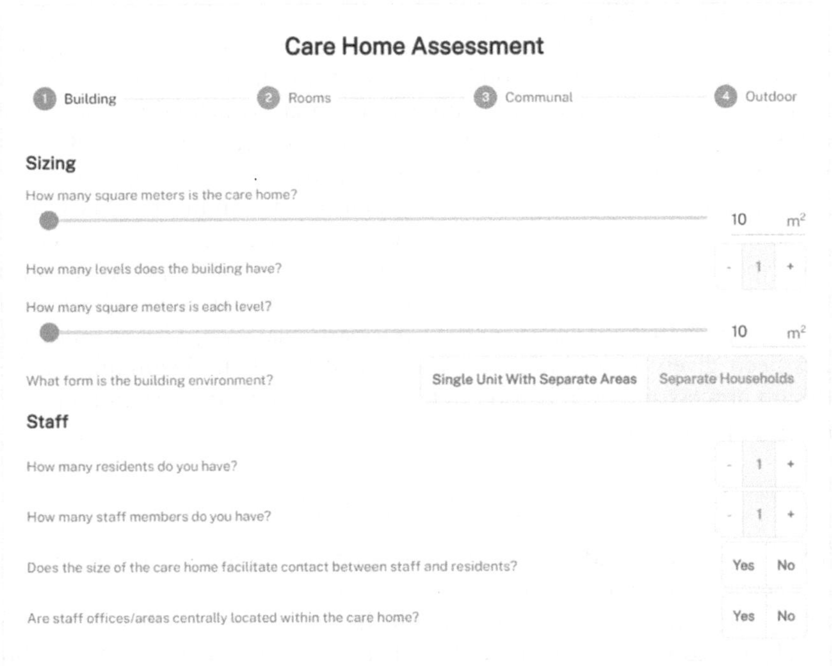

## Care Home Assessment

1 Building     2 Rooms     3 Communal     4 Outdoor

**Sizing**

How many square meters is the care home?

●———————————————————————————— 10   m²

How many levels does the building have?     - 1 +

How many square meters is each level?

●———————————————————————————— 10   m²

What form is the building environment?    Single Unit With Separate Areas    Separate Households

**Staff**

How many residents do you have?     - 1 +

How many staff members do you have?     - 1 +

Does the size of the care home facilitate contact between staff and residents?     Yes   No

Are staff offices/areas centrally located within the care home?     Yes   No

**Fig. 4.3** Screen of the app showing the multiple-answer choice as visualised by users

Developed in a modular way, each question is combined by specifying its type, the way in which it is weighted in the final score (i.e., which factors it contributes to and of how much), and which category it belongs to (Building, Rooms, Communal, Outdoor). The app also specifies how the answer is scored in relation to each action point (i.e., a question can have its maximum score if it is affected by a given action improvement). This allows us to easily tweak the underlying model, adding questions as we go along, and changing the factor and weighting pertaining to each individual survey question. The final score is automatically computed from all available answers, together with each factor and action (Fig. 4.4). The results are then plotted in a bar chart to show the score of each factor, and a radar plot is used to showcase each factor's improvement in relation to the different actions to be undertaken (Fig. 4.5).

We added all questions and scored them uniformly across all different factors. Each factor is in turn scored uniformly to obtain the final rank of a care home. Once combined, all factors are visualised using a radar plot, where users can see a summary of the overall performance of the care home, as well as the individual contributions of each factor on the total score (Fig. 4.5).

Once the overall performance of the building is visualised, users can access the app to see how the building can be improved by following four main actions: (improve on building, on common areas, outdoor spaces and rooms). The results of each improvement are visualised on top of the original performance score, so users can quickly realise which factor has improved and which, in turn, has been negatively affected (Fig. 4.6).

The app has been equipped with the Hotjar system https://www.hotjar.com in order to track how users interact with it and collect feedback on the experience. This allows us to run and track remote experiments with users. In this example, we can see how a rule-based approach informs changes in spatial configurations. In this particular case, a lot of emphasis has been given to people's perception of space at the beginning of the design process and even before, in the establishment of the rules that govern the spatial changes.

## Case Study 2: Magnetising Floor Plans

The second case study has been developed by Foteini Papadopoulou in our research team and explores the automatic spatial arrangement prioritising social and cultural elements in safe houses in the UK (Papadopoulou 2021; Papadopoulou et al. 2022). The algorithms developed for this

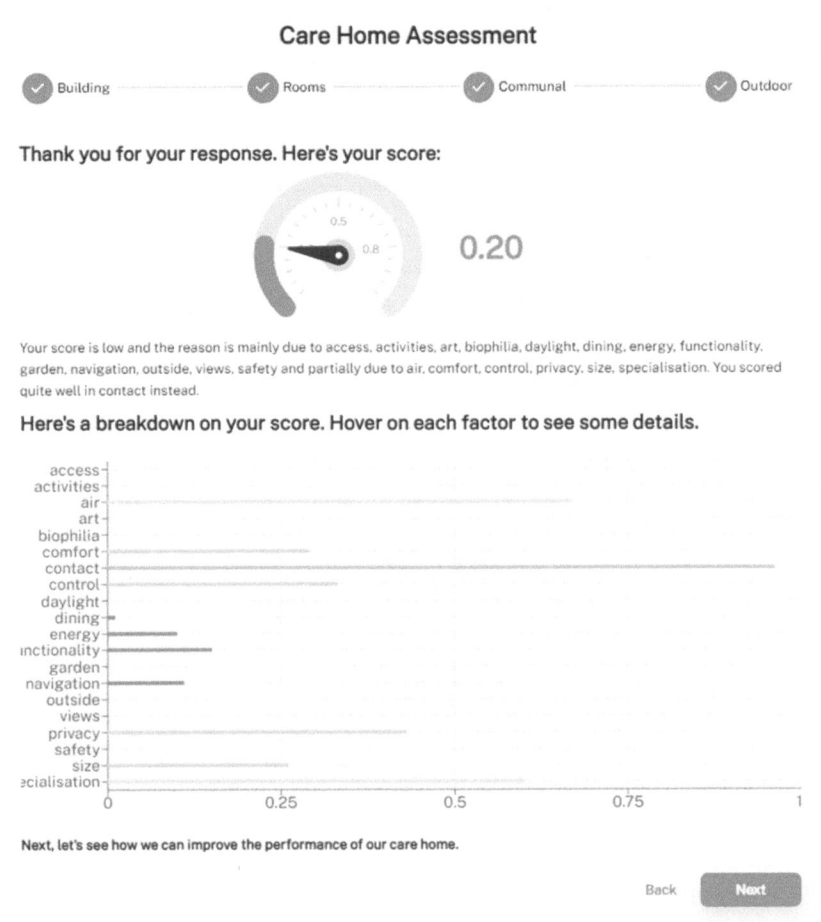

**Fig. 4.4** Visualisation of the performance for each factor

project are based on a number of design criteria that are summarised in the table below (Table 4.1).

After identifying the appropriate design principles, we investigated how to create the best and smartest solutions by combining these principles. We used two different techniques to produce various potential spatial layouts and then compared the outcomes. The adjacency map was created based on the design principles and served as an input for the two methods.

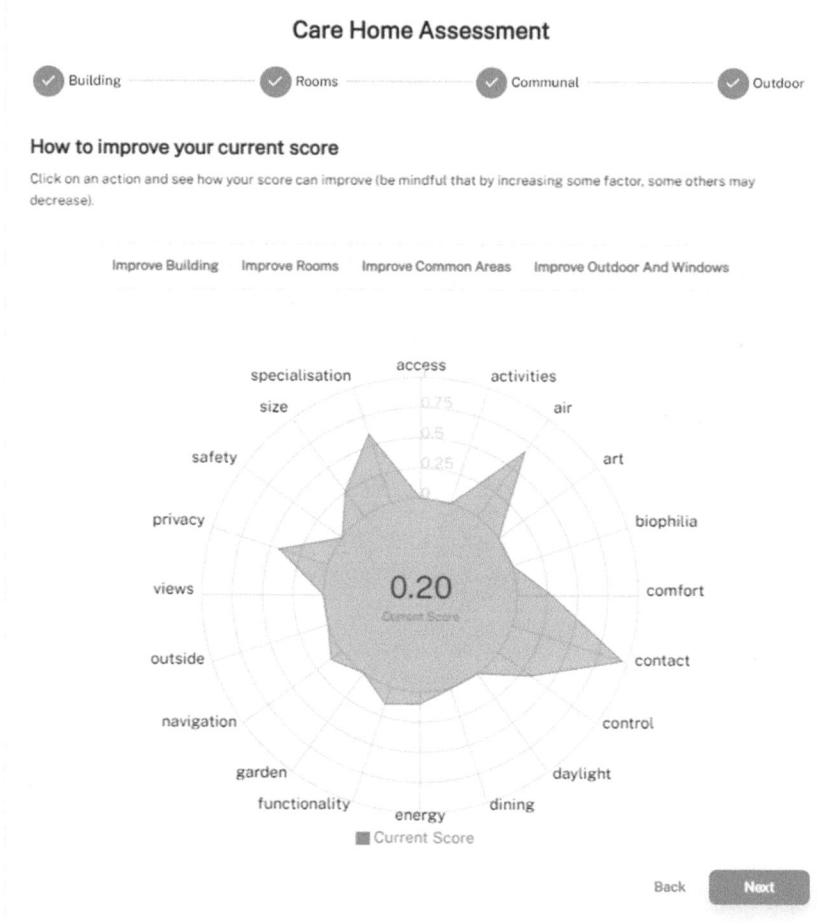

**Fig. 4.5** Summary of factors performance using a radar plot

## *Network of Adjacencies*

Once the list of activities and functions has been created, we organised the list into a hierarchy of functional requirements. This allowed us to create adjacencies between specific spaces based on the social and functional relationships that we want to establish. Therefore, an adjacency network was created to ensure the space functions as intended.

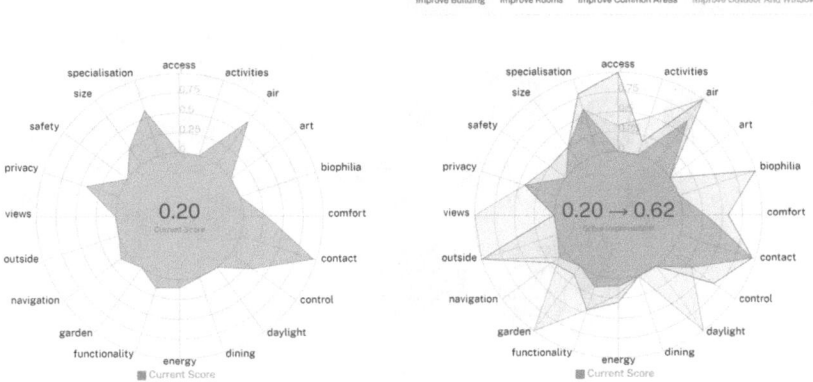

**Fig. 4.6**  Comparison between the original score (left) and the improved performance of the care home (right)

**Table 4.1**  The design principles and the constituent areas of a safe house. Source: Papadopoulou

| Privacy | Outdoor view | Nature Control | Materials Navigation | Facilities & staff |
|---|---|---|---|---|
| Living area | Learning area | Recovering area | | Admin area |
| Bedrooms (en-suite) | Computer training room | Medical care room | | Staff bedrooms |
| Laundry room | Educational training room | Nurse's office | | Staff kitchen |
| Laundry room (staff) | Meeting room | Counselling room | | Manager's office |
| Kitchen | Library | Group meeting room | | Staff office |
| Dining room | Toilets | Arts and crafts room | | Meeting room |
| Living room | | Arts and music therapy room | | Security office |
| TV/cinema room | | Toilets | | Reception area |
| Fitness room | | | | Storage room |
| Garden | | | | Loading area |
| Toilets | | | | Toilets |

**Fig. 4.7** The network of adjacencies of spaces. Activities are here represented as nodes in a graph structure. Different colours indicate different clusters of activities, following the adjacency study. Source: Papadopoulou (2021)

The adjacency map is a useful tool that helps organize different activities within a complex program, highlighting important nodes and the interdependence among them. Additionally, a detailed spreadsheet was generated to outlines the functional needs of each room and establishes a hierarchy of these needs based on design principles and activities that occur in each room. As part of the project, two computational design methods were explored: the Squarified Treemap and the Magnetizing Floor plan Generator (Fig. 4.7).

### The Squarified Treemap

The Squarified Treemap algorithm (Bruls et al. 2000) was created to provide efficient graphical representations of hierarchical information in regular layouts. It is a variation of the Treemap algorithm (Shneiderman 1992) which displays a full hierarchical data structure using adjacent squares based on their relationship in the data tree. The Squarified Treemap algorithm improves on this by approximating areas to squares, even if that means using elongated rectangles for some data. (Marson and Musse 2010) implemented this algorithm in creating floor plans for gaming environments. By using it for spatial configuration, it is possible to generate precise interior layouts with details about their characteristics (Marson and Musse 2010). For this project, we used an implementation of the Squarified Treemap developed by (Holth 2017) based on code created by (Laserson 2021). The design process started by determining the layout parameters derived from the design principles for safe houses.

It is important to see how the design criteria have been created, based on intended functions and an organised list of actions. The actions are described below (Fig. 4.8):

The logic steps of this method are visually summarized in the diagram below (Fig. 4.9).

We can see in Fig. 4.10 how the different rooms and spaces are combined as a result of the squarified treemap method.

The Squarified Treemap method presented two main problems. Firstly, an optimised strategy was employed to arrange the different areas of function and rooms based on an ordered list of areas, starting from the larger ones and progressing to the smaller ones. This strategy focused on filling all the available space into squared areas, thereby resulting in the programmatic functions being arranged solely by size without regard to adjacency. This approach did not comply with the predefined parameters set, such as the design principles and fixed position of specific rooms. To address this issue and ensure adherence to the parameters, the order in which the areas of the rooms were introduced into the algorithm had to be carefully considered and manually rearranged. Secondly, the Squarified algorithm is used to establish the spatial configuration within a specified boundary. However, the algorithm requires a predetermined area for it to function effectively. Consequently, there are limited options for designing a layout that conforms to our design principles.

In order to improve the optmisation process of this method, we implemented a genetic algorithm (GA) to enhance the performance of the Squarified Treemap algorithm. To do that, we followed the workflow described here.

Step 1. A Genetic Algorithm (GA) was employed to identify the optimal placement of bedrooms within the bedroom area concerning the gar-

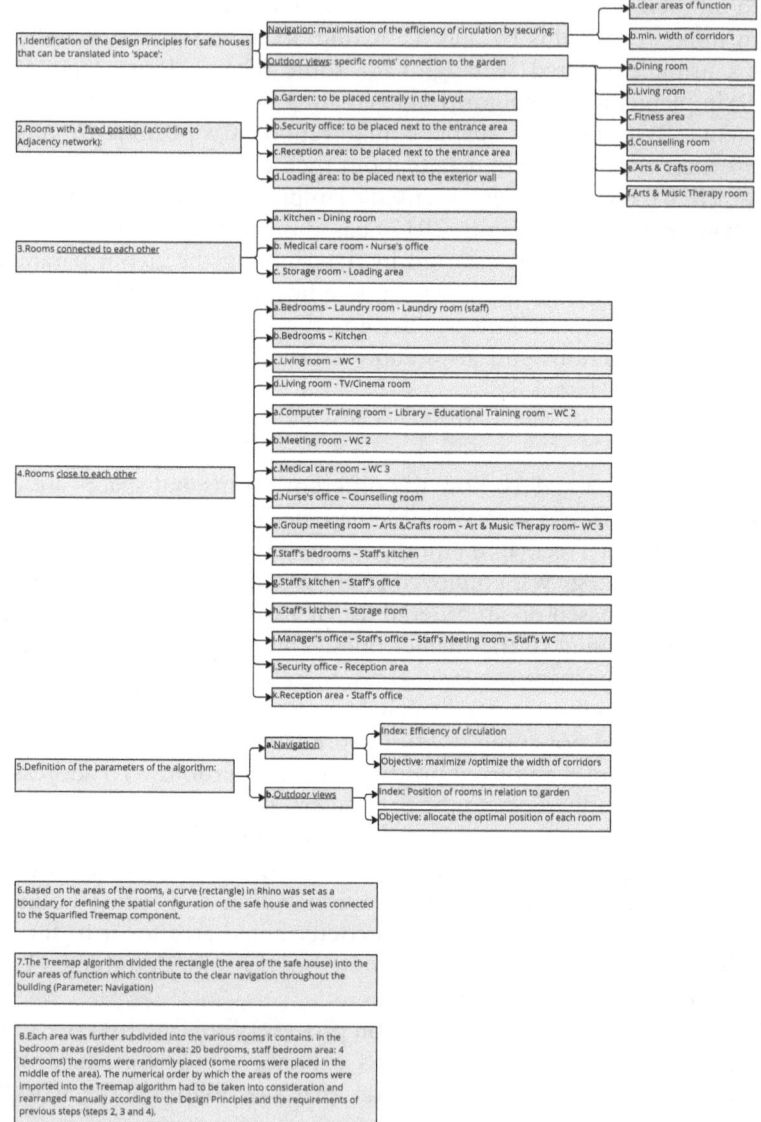

**Fig. 4.8**  Hierarchy of criteria

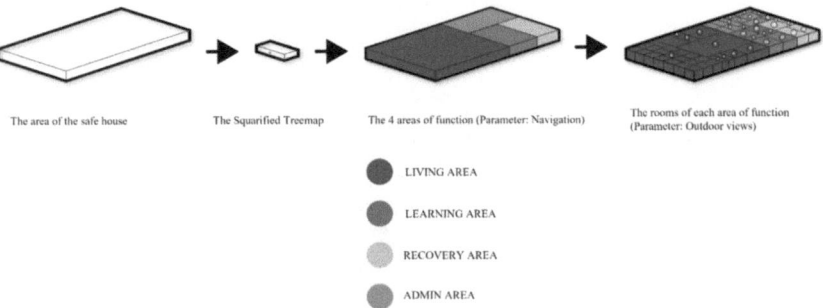

The area of the safe house
The Squarified Treemap
The 4 areas of function (Parameter: Navigation)
The rooms of each area of function
(Parameter: Outdoor views)
LIVING AREA
LEARNING AREA
RECOVERY AREA
ADMIN AREA

**Fig. 4.9** The Squarified Treemap process. Source: Papadopoulou (2021)

1 Bedrooms
2 Laundry room
3 Laundry room (s)
4 Kitchen
5 Dining room
6 Garden
7 Fitness area
8 TV/Cinema room
9 Living room
10 WC1
11 A&C room
12 Group Meeting room
13 Counselling room
14 A&M Therapy room
15 Medical care room
16 WC3
17 Nurse's office
18 Computer room
19 Educational room
20 Library
21 WC2
22 Meeting room
23 Manager's office
24 Meeting room (s)
25 WC(s)
26 Bedrooms (s)
27 Kitchen (s)
28 Storage room
29 Loading area
30 Staff's office
31 Security office
32 Reception area

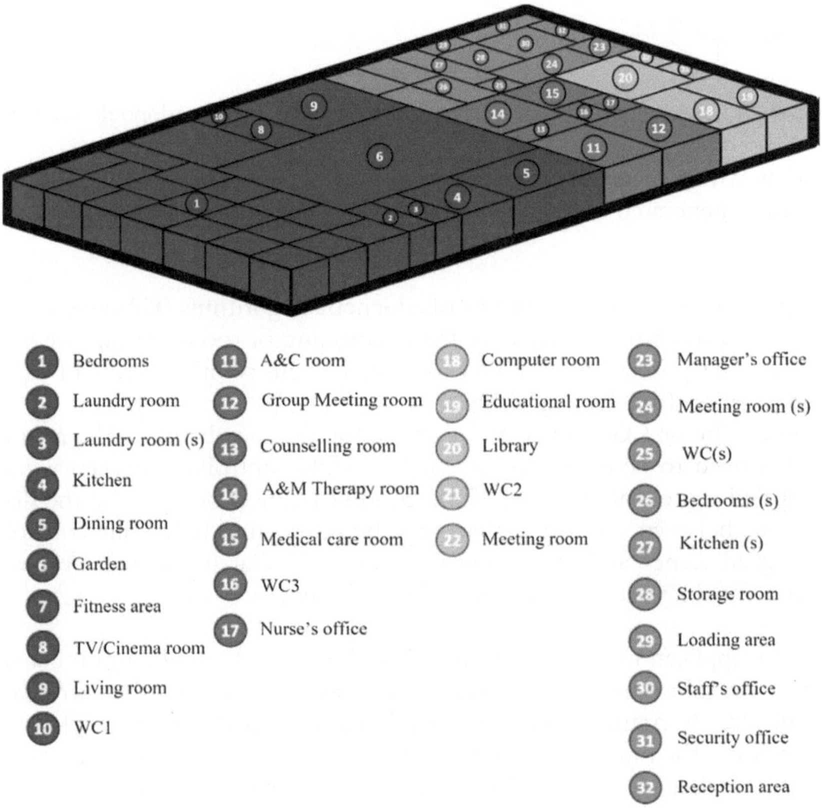

**Fig. 4.10** The rooms within the areas of function. Source: Papadopoulou (2021)

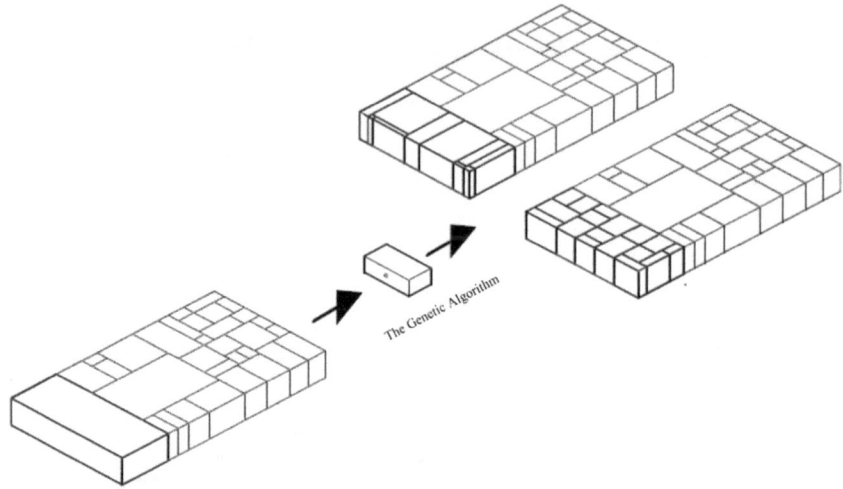

**Fig. 4.11**    GA for optimal placement of bedrooms. Source: Papadopoulou (2021)

den, with a focus on the outdoor view parameter. The GA's implementation generated a variety of solutions, as depicted in Fig. 4.11.

Step 2. Three separate iterations of a Genetic Algorithms (GAs) were utilised in this study to enhance the positioning of rooms within the layout. This process did not alter the initial layout solution created by the
Squarified Treemap algorithm.

Step 3. The first GA aimed to minimise the distance between the rooms
that need to be linked to the garden, while controlling adjacency and
optimising outdoor views. The second GA aimed to minimise the distance between the rooms that must be interconnected, while controlling adjacency and optimising navigation. Similar to the first process,
the initial layout solution remained unaltered.

The application of a GA in steps 2 and 3 yielded a limited number of
improved solutions due to the limited number of rooms, and their arrangement already optimally met the requirements and design principles. In
such cases, the numerical order by which the size of the rooms was listed
in the Treemap Algorithm was manually rearranged. Conversely, when the

GA was applied in step 1 to determine the optimal location of bedrooms with minimal distance to the garden, several solutions were generated, as illustrated in Fig. 4.11. However, some of the solutions produced odd-shaped rooms as the GA algorithm optimises room shapes based on their sizes listed in the algorithm. We then moved to the application of the second self-organising method using the Magnetising Floor Plan Generator.

### The Magnetising Floor Plan Generator (MFG)

The algorithm developed by (Egor et al. 2020) is specifically designed to generate spatial configurations for public buildings by taking into account the required adjacencies between interior spaces. The algorithm is user-friendly, and it allows users to input the fundamental characteristics of each room, and then it calculates the required corridors around each room. Depending on the options chosen within various parameters, different outcomes can be generated, from which the optimal solution is chosen based on the evaluation function, as described by (Bielik 2019; Hellguz 2021).

As in the previous iteration, we rationalised the design goals through an organised list of actions. To establish a functional layout, the first step was to create a network of adjacencies between rooms based on the Adjacency network (Fig. 4.7) while importing the name and area value of each room. By adding two more links between the kitchen-storage room and living room-library, all functional areas were connected (see Fig. 4.12).

Next, one room was designated as an entrance (reception area). It was necessary to ensure that the location of the reception area and the security office were adjacent to the entrance area (as shown in Fig. 4.12).

To develop the layout of the building, a curve (rectangle) was drawn, which established the boundaries within which the layout was to be developed. These elements were also connected to the MFG algorithm to further aid in the development of the layout.

The algorithm utilized three parameters, namely the number of iterations, maximum distance between rooms, and the type of corridors, to establish connections between rooms. Different layouts were generated as previews based on these parameters. After multiple attempts, the exploration of this process resulted in the successful placement of all 32 rooms of the safe house. As a result, 7 layouts were produced (as shown in Fig. 4.13).

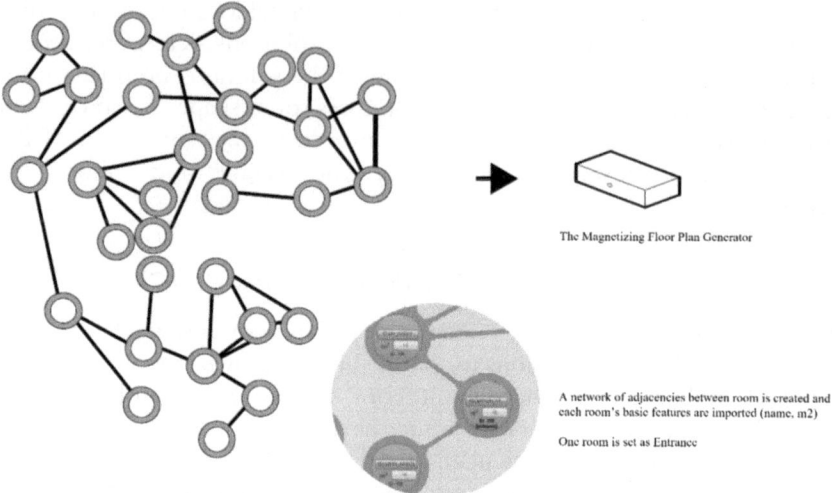

**Fig. 4.12** The Magnetising Floor Plan Generator process. Source: Papadopoulou (2021)

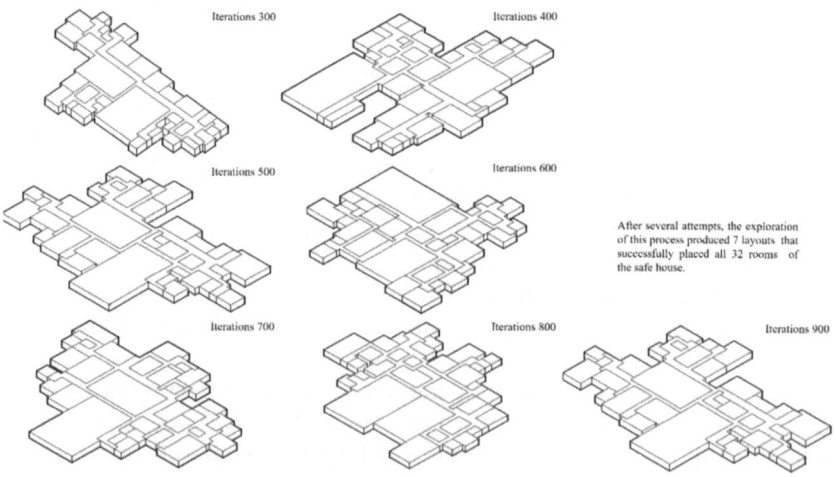

**Fig. 4.13** The layouts with iteration 300, 400, 500, 600, 700, 800, 900. Source: Papadopoulou (2021)

### Final Layout

After numerous generations, the layout produced with iteration no. 700 was selected as the optimal spatial configuration. Generation number 700 met all intended criteria for safe house design elaborated as the premise for this study.

- The four areas of function (living, recovery, learning, and administrative) are clearly defined, enabling efficient circulation of residents.
- Residents and staff areas are clearly defined, enhancing the sense of safety, security, and isolation for women during their stay and upon arrival.
- The garden is centrally placed for maximum security and privacy.
- The garden's position maximizes the number of rooms with a view of it and ensures a connection between specific rooms and the garden.
- The placement of the rooms with fixed positions is secured, as required by the Adjacency network, including the garden (centrally placed), security office, reception area (next to the entrance area), and loading area (next to the exterior wall).
- The required adjacencies between specific rooms are ensured.

Figure 4.14 shows the selected layout.

## Case Study 3: Self-organising Floor Plans in Care Homes

This third project discusses a study we carried out in 2020 (Carta et al. 2020) where we used neural networks to generate optimised spatial configurations.

### Evolutionary Approaches to Design

This project presents a method for enhancing the floor plan efficiency in care homes by using a self-organising genetic algorithm. The goal is to reduce energy consumption, improve the wellbeing of residents and have a positive impact on the costs of energy and health care. In order to achieve the optimal spatial configuration, several design criteria were developed and tested based on existing literature reviews and initial considerations of care home layouts. These criteria are used as objectives in a Genetic Algorithm (GA) to evaluate the best design solution. The self-organised

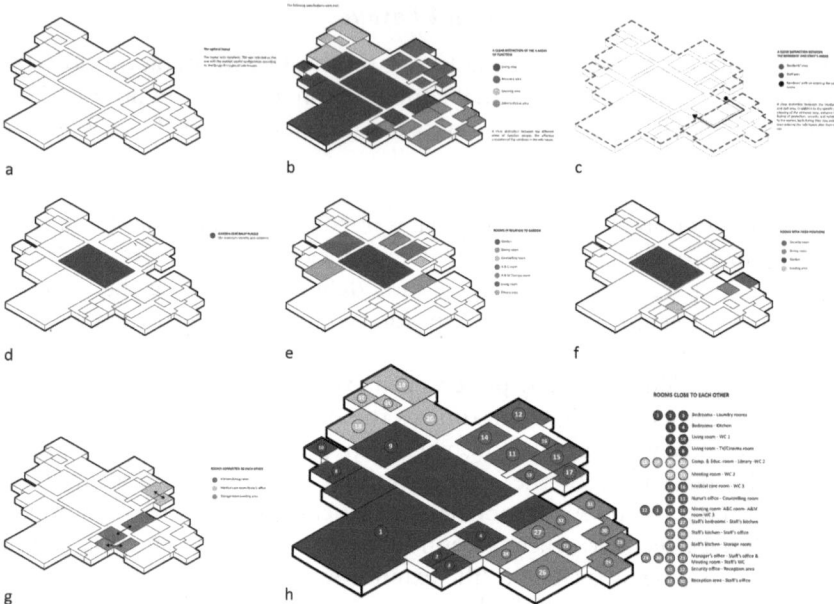

**Fig. 4.14** The optimal layout (Iteration: 700). (**a**): General layout, (**b**): clear distinction between the four areas of function, (**c**): clear distinction between the residents and staff areas, (**d**): centrally positioned garden, (**e**): rooms connections to the garden, (**f**) rooms with fixed positions, (**g**): rooms connected to each other, (**h**) rooms close to each other. Source: Papadopoulou (2021)

floor plan is then used to run a final simulation to observe how residents could use the optimised spaces and to measure the improved efficiency of the new plans.

With this project, we contend that the efficiency of building floor plans has a direct impact on the three pillars of sustainability: environmental, social and economic. Accordingly, the primary focus of this study is to propose a method to optimise the design of spatial layouts in care homes based on predetermined design criteria. We suggest that this method may help designers to create more sustainable and energy-efficient floor plans in care home projects. The design criteria used in this study are based on existing literature on spatial characteristics of layouts in facilities with people with dementia.

Traditional algorithms used for optimising spatial solutions for floor plans employ a brute-force approach of generating all possible solutions and selecting the one that best fits the given requirements. These techniques require intensive computations and use of computing resources (Kleinberg and Tardos 2006). In contrast, this study employs Genetic Algorithms (GAs), which are designed to address problems where multiple objectives need to be computed simultaneously (Mitchell 1998), GAs are analogous to the natural selection process where the highest performing breeds have a better chance of survival and lead to the evolution of a certain species. The best matches are selected to find an optimised chromosomal combination.

GAs have been extensively employed in design to find optimised spatial solutions for various tasks (Miles et al. 2001) including spatial arrangement (Caldas and Norford 2002), location of specific activities in a building (Liu et al. 2003), building shapes (Ouarghi and Krarti 2006; Tuhus-Dubrow and Krarti 2010), and material distribution for energy performance-related tasks (Wang et al. 2018). However, GAs have been less used in the optimisation of spatial layout and spatial planning. In this study, we employed GAs to find an optimised configuration of a floor plan based on design recommendations for residents with dementia.

### *The Workflow*

It is important to note that this study does not aim to identify spatial solutions specifically for people with dementia. To do so would require further studies on the relationship between spatial layout and the various stages of dementia, as well as different needs over time of residents, and the relationships of residents with care workers, visitors, and other people involved in their care and the maintenance of the facility. All these aspects are beyond the scope of this study. The paper concludes with a discussion of the results and suggestions for future studies and experiments using emergence behaviour models to improve sustainable development in design.

To investigate the optimal spatial distribution and general layout design in care homes, a new set of design criteria was developed and the following workflow was employed:

1. Use an existing floor plan of a care home designed using current design criteria as a starting point for the case study (in this case, the Irene Baron Eden Centre, Winnipeg, Manitoba, MMP Architects 2009).

2. Elaborate a new set of design criteria based on existing literature and their interpretation.
3. Design a Genetic Algorithm (GA) using NeuroEvolution of Augmenting Topologies (NEAT) that uses the design criteria (2) as objectives.
4. Run (3) until finding a satisfying value of the fitness function and isolate the corresponding floor plan configuration.
5. Evaluate the efficiency and accuracy of the self-organising approach using the embedded fitness function.
6. Build a model of the optimised configuration (4) using a 3D modelling software and run a BOID simulation (Reynolds 1987) to observe how residents will use the new configuration.
7. Compare the results of the simulation run on the optimised floor plan (6) with the initial configuration of the care home (1).
8. Discuss the results.

To implement the genetic evolution analogy, we started with an initial population of candidate design solutions that were selected randomly in the first iteration and then gradually optimised through a cycle of iterations. This process is depicted in Fig. 4.15.

A genetic algorithm is a computational method that mimics the process of natural selection in order to find an optimal solution to a problem. It works by creating a population of potential solutions, evaluating their fitness based on a pre-defined function, and using selection, mutation, and recombination to create new generations of solutions that are better adapted to the problem at hand (Fig. 4.15). Each iteration of the algorithm creates a new set of solutions called a "generation," which is evaluated using a fitness function that measures how well each solution meets the objectives of the design problem. The most fit solutions are then selected to "evolve" into the next generation. This process involves modifying the solutions by recombining them and randomly mutating them, in order to create a new set of solutions that may be better adapted to the problem. By combining the best performing parts of different solutions, the algorithm creates a new, potentially better solution in each iteration. The algorithm continues to produce new generations of solutions until it either reaches a set number of generations or a population with a satisfactory fitness level. This ensures that the algorithm eventually converges on a solution that meets the design objectives.

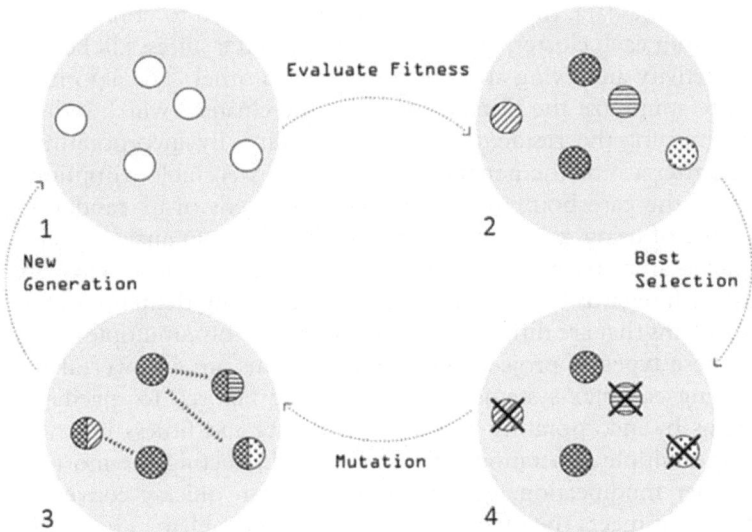

**Fig. 4.15** The genetic algorithm involves a series of steps to generate optimised solutions. Firstly, at each iteration, candidate solutions are generated and evaluated based on a fitness function (1). Then, the best solutions are selected and used as new breeders, while the worst ones are removed (2). Subsequently, the best solutions undergo random mutations to generate new candidate functions (3). Finally, the new candidate functions are inserted into a new generation for further evaluation and optimisation (4). The genetic algorithm thus uses an iterative approach to find the best solution, with each iteration building on the solutions of the previous one

The graph reflects specific design requirements into the genetic algorithm. After modelling a generic configuration of a care home as a graph, we added nodes to the graph that represent the different rooms and areas within the care home. Twelve nodes were added to represent the suites on each unit, two nodes were added for the dining area and kitchen, and two nodes were added for the activity area and living area. Other rooms found in the case study, such as the salon, spa, and laundry, were added as individual nodes outside of the existing clusters. Each node in the graph stores information about the corresponding room's size in square metres, as well as its adjacency requirements. In this case, we also included the direct proximity of the kitchen to the dining area in the adjacency requirements.

Using these data, the genetic algorithm attempted to minimize the distance within each cluster, ensuring that each unit's suites, kitchen, dining room, activity and living area are close to each other. The algorithm also aimed to minimize the distance between the clusters, which reduces the travel time for the residents within the facility. By incorporating these design criteria into the genetic algorithm, we were able to optimise the layout of the care home to better meet the needs of its residents. One advantage of using genetic algorithms is that they can analyse a large and complex search space in a relatively short amount of time. They are also robust with regards to noise and uncertainty, making them useful for solving problems that are difficult to model or that involve multiple objectives.

In these types of projects, genetic algorithms are a powerful tool for optimising complex systems, and they can be tailored to specific design problems by incorporating relevant constraints and fitness functions. By creating multiple generations of solutions and selecting the most fit ones for further modification, genetic algorithms can quickly converge on a solution that meets the objectives of the design problem. The reader may want to substitute rooms with users and spatial requirements with social characteristics within the graph structure to understand how such complex networks can be manipulated to find optimal configurations based on given objectives, as described in Chap. 2.

### *Evolving Spatial Configurations*

In this study, we utilised a genetic algorithm (GA) strategy based on the NeuroEvolution of Augmenting Topologies (NEAT), which was initially developed by Joel Simon (Stanley and Miikkulainen 2002; Simon 2020). To apply the GA strategy to optimise floor plan designs, we needed to encode the solutions (i.e., the floor plans) and specify the evolutionary processes, such as the selection of the most fit solutions within each generation and their mutation into a new generation.

NEAT is a method that effectively represents topologies to be evolved using a GA. Each solution (or genome) is represented by a series of nodes and connections. Nodes store information used by the fitness function to evaluate the corresponding solution, while connections store the linked nodes, their weight (usually representing the cost of that connection, e.g., the distance between the rooms), and whether it is enabled or not in the current instance. The method allows for new connections between existing nodes or the addition of new nodes to the graph during the mutation of each solution.

After a random mutation takes place, the graph is converted into a candidate solution and added to the new generation that will be evaluated through the chosen fitness function. The work of Stanley and Miikkulainen (Stanley and Miikkulainen 2002) provides a clear illustration of this method, as seen in the graph solution they presented in (Stanley and Miikkulainen 2002:106–107). In this manner, the mutation process of each solution within the GA can potentially result in the formation of new connections between nodes that already exist, or the addition of entirely new nodes to the graph structure.

After a random mutation has occurred in the genetic algorithm, the resulting graph is transformed into a candidate solution and included in the new generation. This new generation of solutions is then evaluated using a fitness function that has been chosen beforehand. The NEAT method provides a means of encoding the problem of optimising floor plan designs and devising a genetic algorithm that can successfully address it. As such, the GA strategy that was implemented for the purpose of evolving floor plans can be summarised as follows:

1. Create an initial population of random graphs that connect the rooms while meeting adjacency requirements, if applicable.
2. Apply NEAT mutations to create a new genome.
3. Translate the new genome into corresponding floor plan designs.
4. Utilize an Ant-Colony Optimisation algorithm to create hallways that connect the rooms.
5. Assess candidate solutions using a predetermined fitness function.
6. Convert candidate solutions into a new generation.
7. Repeat the process starting from step (2) until a specified number of generations have been evaluated.

The mapping process in step 3 of the floor plan optimisation procedure comprises of three stages. Firstly, the nodes are arranged using a spectral layout technique. Secondly, a physics simulation algorithm is employed to position the nodes to their final coordinates, which correspond to the centre of the rooms in the floor plan. Lastly, the layout is transformed into a polygonal mesh that accurately represents the walls (Fig. 4.16).

The process of mapping graphs generated by the NEAT's mutations into a floor plan design is followed by the generation of hallways using an Ant-Colony Optimisation (ACO) algorithm (Dorigo and Gambardella 1997). ACO is a probabilistic approach that seeks to find optimal paths

**Fig. 4.16** The three steps (**a–c**), and (**d**) of the process mapping a graph into a floor plan design. Source: Simon (2020)

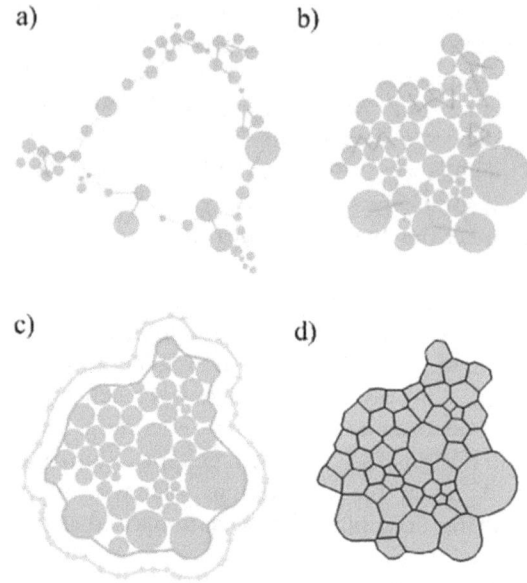

through graphs, and is inspired by the foraging behaviour of ants. Ants explore their environment randomly, but when they find food, they return to the colony and deposit pheromones. Other ants follow the pheromone trail, reinforcing the path. This creates a signal that guides the ants between the food source and the colony. In the ACO algorithm used in this study, hallways are generated by creating a graph connecting the nodes representing the rooms in the floor plan, and then simulating the behaviour of ants to find the optimal paths.

The process of generating hallways along the edges of the resulting polygonal mesh using the ACO algorithm is illustrated in Fig. 4.17 for a three-room floor plan (Fig. 4.17a). First, interior nodes and edges are created for each room (Fig. 4.17b). Then, an ACO strategy is employed to find the optimal path connecting the rooms (Fig. 4.17c). Finally, the hallways are smoothed by moving their vertices halfway to their neighbours (Fig. 4.17d).

After the hallway generation process, the resulting hallways are added to the floor plan design that was previously generated without them. This produces a new floor plan design that includes the hallways (Fig. 4.18a). Finally, the interior edges of the hallway polygonal mesh are used to

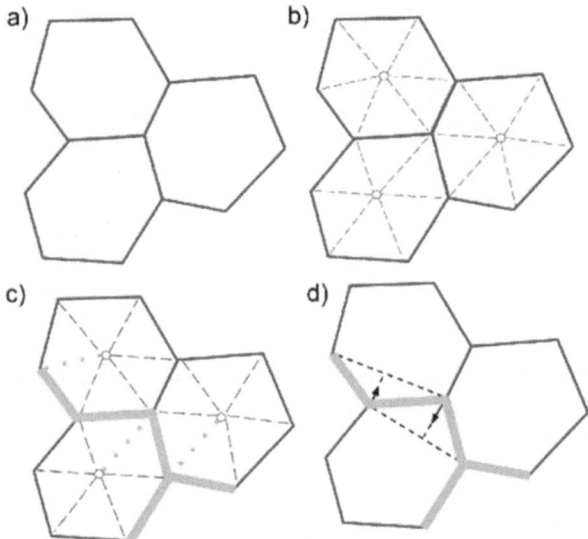

**Fig. 4.17**   An overview of the hallway generation strategy based on Ant-Colony Optimisation. Source: Simon (2020). The figure illustrates the process of hallway generation in a generic configuration of 3 cells (**a**), where nodes and connections are generated inside each cell (**b**), the ACO method is used to find optimised connections among cells (**c**), and their optimisation (**d**)

determine door placements, and the hallways are merged with the rest of the floor plan to produce a final geometry (Fig. 4.18b). This final step completes the process of generating a floor plan design that satisfies the given requirements and is optimised using a GA strategy with NEAT mutations and an ACO algorithm.

In step 5, the fitness function used for evaluating the candidate solutions is highly dependent on the specific optimisation objectives of the initial problem. For this particular study, the aim was to generate an optimal floor plan design based on several criteria established for this project. The selection of the most appropriate criteria to be included in the NEAT encoding was carefully made, based on the team's judgement of what aspects were better captured by the optimisation strategy. The final result of the optimisation process is presented in Fig. 4.19, while the corresponding fitness function diagram is shown in Fig. 4.20.

a)                                    b)

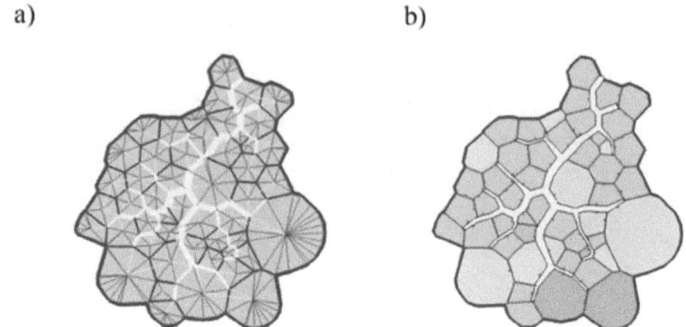

**Fig. 4.18** The final floor plan with the generated hallways: (**a**) shows the application of the hallway finding method to an entire floor plan, and final geometry (**b**). Source: Simon (2020)

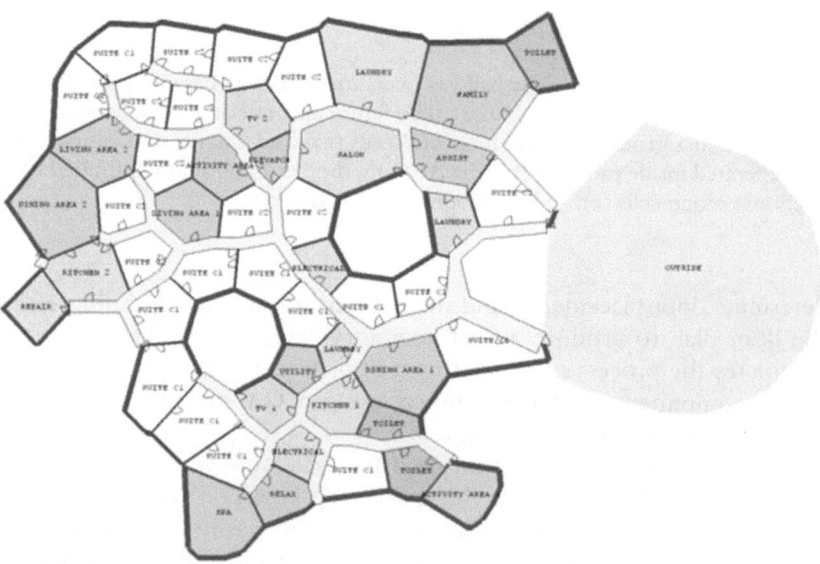

**Fig. 4.19** The final floor plan with the optimised spatial configuration

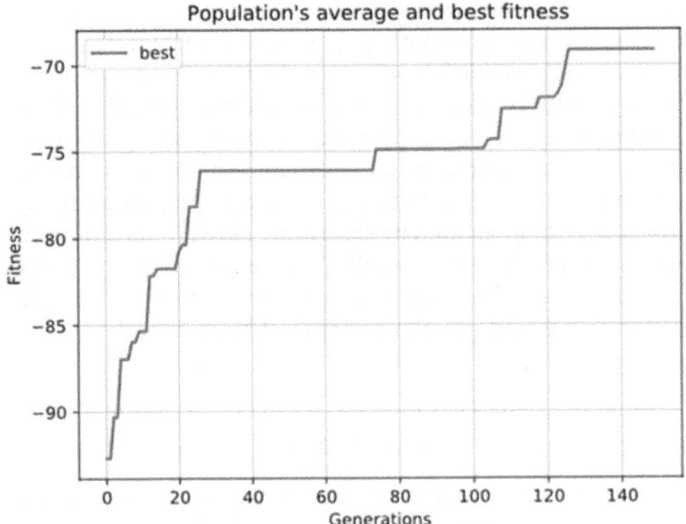

**Fig. 4.20** Fitness function diagram, showing best fit solutions after 120 generations

This third case study focuses on developing a workflow to optimise bidimensional spatial configurations (floor plans) in care home settings. This work is part of a larger project aimed at testing the principles of automatic optimisation using genetic algorithms (GA). The study also aims to explore how design criteria from existing studies can be used to generate inputs for the GA.

The results of the study indicate that inputting a limited set of design criteria into the GA can produce optimised configurations that can be quantitatively measured. However, the criteria should be expanded to include functional adjacencies and a broader range of design recommendations. For example, the resident suites in the generated layout are distributed rather than clustered together in each unit. While clustering is typically used to achieve quieter zones within the pod and reduce intrusions from passing residents, other solutions like set-back doorways and acoustic wall and door construction specifications can be considered. Moreover, the distributed spatial organization generated by the GA may allow for planned infection control of residents within areas of the care home with access to amenities.

The present study demonstrates that GAs can be a useful tool for analysing and designing care home building layouts to generate more sustainable spatial configurations. They can consider a wider range of approaches to spatial organisation than traditional design practices. Modelling simulations of resident activity within generated spatial organisations can allow for empirical analysis and testing of relevant criteria like visual access and circulation paths, leading to better configurations. This last project illustrates how space can generated by algorithms. Private and common rooms, corridors, entrances and the overall spatial experience is not designed in the traditional sense of architecture; it is rather the result of a number of computations governed by rules established by designers.

## CONCLUSIONS: GENERATION OF DATA-DRIVEN PUBLIC SPACE

This chapter illustrates three different cases where spatial configurations are the result of instructions given through code. In the very first example, we saw how we can give clear instructions, dimensions and maximum area constraints and ask a very simple script to generate a rudimentary floor plan for us (Fig. 4.1). With AISLA we can then see how we can work with specific design criteria to manipulate the spatial configurations of buildings, in that particular case of care home facilities. This project is helpful to illustrate how we can attribute specific importance to certain criteria over the others by assigning different weights to different factors. The designer or the facility manager is then able to visualise and understand trade-offs between design goals and find the preferred solution. Trade-offs may include better spatial configuration, less production of $CO_2$, or reduction of energy consumption versus costs or temporary disruption during the work.

In the second case study, magnetising floor plans we can see how some of the decision-making process is shifted from the designer/manager to a computation. The designer sets up initial rules and objectives and a solver would yield a spatial configuration as an output of the computation. The task of the algorithm is here to heuristically try different solutions until no further improvement can be made. The designer sets up an objective and the algorithm needs to achieve spatial configurations that are the closest possible to the objective.

The final case study, self-organising floor plans shows how this computation can be achieved using more sophisticated methods, in this case through generative approaches and genetic algorithms. This is the only case in the examples included in this chapter where the algorithms work as a black box. This means that the calculations reach some point where the inner complexity is too high for the human brain to comprehend. The designer sets the rules and defines the characteristics of the expected results. In this way they can assess and measure the extent to which the computation has been successful or not. However, it is impossible for anyone to follow step-by-step how the algorithm arrived at a specific result, due to the complexity of the inner mechanisms underpinning the calculation. It is important to note that in the space of computational design, social and cultural aspects of the resulting spatial configurations are generally considered by designers during the elaboration of the underpinning criteria and objectives. Unlike a traditional design process, where such aspects are either inherently considered by the designer as a part of their initial conceptual schemes and subsequent refinements, in computational approaches social aspects of space are encoded within the process. This means that social questions require to be translated into a form that can be calculated within the script. Most of the time this translation is from qualitative aspects to a measurable set of features. While in the traditional design approach to spatial configurations the social aspects of space are included as assumptions by the designer and inextricably connected to the initial shapes of the building, in computational design such aspects are codified and measured within the computer programme that will yield the final space.

It is often, in the process of encoding cultural and social aspects that design may implicitly work within a certain degree of bias in the process in the form of design assumptions. This aspect if further discussed in the following chapter.

## NOTES

Part of the literature review included in this chapter has been developed in (Carta 2021).

The first case study has been developed as a part of a project on automation and spatial improvement of care home design funded by Research England through the University of Hertfordshire.

Case Study 2 Magnetising Floor Plans has been published in various forms in Papadopoulou F. (2021) *The development of a functional accommodation for female victims of sex trafficking: How its design will ensure the implementation of the programmes of assistance and the covering of their needs, leading to their practical and psychological support that will help them adjust and adapt back to 'normality'*. University of Hertfordshire, Hatfield and Papadopoulou F, Carta S and Owen IW (2022) Safe Houses: design principles, potentials and limitations. An analysis through data-driven approaches. In: *10th Annual International Conference on Industrial, Systems and Design Engineering*, Athens, Greece, 20 June 2022. ATINER.

Part of Case Study 3 Self-Organising Floor Plans in Care Homes has been published in Carta S, St. Loe S, Turchi T, et al. (2020) Self-organising floor plans in care homes. *Sustainability* 12(11). MDPI: 4393.

## REFERENCES

Anderson C, Bailey C, Heumann A, et al. (2018) Augmented space planning: Using procedural generation to automate desk layouts. *International Journal of Architectural Computing* 16(2). SAGE Publications Sage UK: London, England: 164–177.

Bielik M (2019) Magnetising Floor Plan Generator. Available at: https://toolbox. decodingspaces.net/magnetizing-floor-plan-generator/.

Brookes MJ and Kaplan A (1972) The office environment: Space planning and affective behavior. *Human factors* 14(5). SAGE Publications Sage CA: Los Angeles, CA: 373–391.

Bruls M, Huizing K and Van Wijk JJ (2000) Squarified treemaps. In: *Data Visualization 2000: Proceedings of the Joint EUROGRAPHICS and IEEE TCVG Symposium on Visualization in Amsterdam, The Netherlands, May 29–30, 2000*, 2000, pp. 33–42. Springer.

Caldas LG and Norford LK (2002) A design optimization tool based on a genetic algorithm. *Automation in construction* 11(2). Elsevier: 173–184.

Carta S (2021) Self-Organizing Floor Plans. *Harvard Data Science Review HDSR*. Epub ahead of print 2021.

Carta S, St. Loe S, Turchi T, et al. (2020) Self-organising floor plans in care homes. *Sustainability* 12(11). MDPI: 4393.

Chen Y, Zhang C, Qiao T, et al. (2021) Ship detection in optical sensing images based on YOLOv5. In: *Twelfth International Conference on Graphics and Image Processing (ICGIP 2020)*, 2021, pp. 102–106. SPIE.

Dorigo M and Gambardella LM (1997) Ant colony system: a cooperative learning approach to the traveling salesman problem. *IEEE Transactions on evolutionary computation* 1(1). IEEE: 53–66.

Eastman CN (1975) *Spatial Synthesis in Computer-Aided Building Design*. Elsevier Science Inc.

Egor G, Sven S, Martin D, et al. (2020) Computer-aided approach to public buildings floor plan generation. Magnetizing Floor Plan Generator. *Procedia Manufacturing* 44. Elsevier: 132–139.

Guo Z and Li B (2017) Evolutionary approach for spatial architecture layout design enhanced by an agent-based topology finding system. *Frontiers of Architectural Research* 6(1). Elsevier: 53–62.

Hellguz (2021) Magnetizing Floor Plan Generator. Available at: www.food4rhino.com/en/app/magnetizing-floor-plan-generator.

Holland JH (1996) *Hidden Order: How Adaptation Builds Complexity*. Addison Wesley Longman Publishing Co., Inc.

Holland JH (2000) *Emergence: From Chaos to Order*. OUP Oxford.

Holland JH (2002) Complex adaptive systems and spontaneous emergence. In: *Complexity and Industrial Clusters: Dynamics and Models in Theory and Practice*, 2002, pp. 25–34. Springer.

Holth T (2017) Treemap. Available at: Available from: Grasshopper3d.com https://www.gras 48 shopper3d.com/forum/topics/treemap (accessed 6 April 2022).

Jo JH and Gero JS (1998) Space layout planning using an evolutionary approach. *Artificial intelligence in Engineering* 12(3). Elsevier: 149–162.

Johnson, S., 2002. Emergence: The connected lives of ants, brains, cities, and software. Simon and Schuster. New York.

Kleinberg J and Tardos E (2006) *Algorithm Design*. Pearson Education India.

Laserson U (2021) Squarify. Available at: https://github.com/laserson/squarify.

Levary RR and Kalchik S (1985) Facilities layout—A survey of solution procedures. *Computers & Industrial Engineering* 9(2). Elsevier: 141–148.

Liggett RS (2000) Automated facilities layout: past, present and future. *Automation in construction* 9(2). Elsevier: 197–215.

Liu DK, Yang YL and Li QS (2003) Optimum positioning of actuators in tall buildings using genetic algorithm. *Computers & structures* 81(32). Elsevier: 2823–2827.

March L and Steadman P (1971) *Spatial Allocation Procedures. In The Geometry of Environment*. MIT Press.

Marson F and Musse SR (2010) Automatic real-time generation of floor plans based on squarified treemaps algorithm. *International Journal of Computer Games Technology* 2010. Hindawi Limited London, UK, United Kingdom: 1–10.

Martin J (2005) Algorithmic beauty of buildings methods for procedural building generation. *Computer Science Honors Theses*: 4.

Merrell P, Schkufza E and Koltun V (2010) Computer-generated residential building layouts. In: *ACM SIGGRAPH Asia 2010 Papers*, pp. 1–12.

Michalek J, Choudhary R and Papalambros P (2002) Architectural layout design optimization. *Engineering optimization* 34(5). Taylor & Francis: 461–484.

Miles JC, Sisk GM and Moore CJ (2001) The conceptual design of commercial buildings using a genetic algorithm. *Computers & Structures* 79(17). Elsevier: 1583–1592.

Mitchell M (1998) *An Introduction to Genetic Algorithms.* MIT press.

Ouarghi R and Krarti M (2006) Building Shape Optimization Using Neural Network and Genetic Algorithm Approach. *Ashrae transactions* 112(1).

Papadopoulou F (2021) *The development of a functional accommodation for female victims of sex trafficking: How its design will ensure the implementation of the programmes of assistance and the covering of their needs, leading to their practical and psychological support that will help them adjust and adapt back to 'normality'.* University of Hertfordshire, Hatfield.

Papadopoulou F, Carta S and Owen IW (2022) Safe Houses: design principles, potentials and limitations. An analysis through data-driven approaches. In: *10th Annual International Conference on Industrial, Systems and Design Engineering,* Athens, Greece, 20 June 2022. ATINER.

Park H, Suh H, Kim J, et al. (2023) Floor plan recommendation system using graph neural network with spatial relationship dataset. *Journal of Building Engineering* 71. Elsevier: 106378.

Reynolds CW (1987) Flocks, herds and schools: A distributed behavioral model. In: *Proceedings of the 14th annual conference on Computer graphics and interactive techniques,* 1987, pp. 25–34.

Seehof J, Evans W, Friederichs J, et al. (1966) Automated facilities layout program. *Communications of the ACM* 9(7). ACM New York, NY, USA: 478.

Shaviv E (1987) Generative and evaluative CAAD tools for spatial allocation problem. In: *Principles of Computer-Aided Design: Computability of Design,* pp. 191–212.

Shekhawat K (2023) A theory of L-shaped floor-plans. *Theoretical Computer Science* 942. Elsevier: 57–92.

Shneiderman B (1992) Tree visualization with tree-maps: 2-d space-filling approach. *ACM Transactions on graphics (TOG)* 11(1). ACM New York, NY, USA: 92–99.

Simon J (2020) Evolving Floorplans. Available at: https://www.joelsimon.net/evo_floorplans.html.

Stanley KO and Miikkulainen R (2002) Evolving neural networks through augmenting topologies. *Evolutionary computation* 10(2). MIT Press: 99–127.

Tuhus-Dubrow D and Krarti M (2010) Genetic-algorithm based approach to optimize building envelope design for residential buildings. *Building and environment* 45(7). Elsevier: 1574–1581.

Veloso P and Krishnamurti R (2021) Self-learning agents for spatial synthesis. In: *Formal Methods in Architecture: Proceedings of the 5th International Symposium on Formal Methods in Architecture (5FMA), Lisbon 2020,* 2021, pp. 265–276. Springer.

Wang Y, Shen J, Xiang W, et al. (2018) Identifying characteristics of resilient urban communities through a case study method. *Journal of Urban Management* 7(3). Elsevier: 141–151.

Wu W, Fu X-M, Tang R, et al. (2019) Data-driven interior plan generation for residential buildings. *ACM Transactions on Graphics (TOG)* 38(6). ACM New York, NY, USA: 1–12.

Zheng H and Yue Ren (2020) Architectural layout design through simulated annealing algorithm. In: *25th International Conference of the Association for Computer-Aided Architectural Design Research in Asia (CAADRIA 2020): RE: Anthropocene, Design in the Age of Humans,* 2020, pp. 275–284. The Association for Computer-Aided Architectural Design Research in Asia....

# CHAPTER 5

# Ethics, Fairness and Bias in Accidental Collectives

**Abstract** This chapter explores aspects of ethics, fairness and social justice that relate these spaces to accidental collectives. The first section investigates the ethical consequences of accidental collectives' development by critically examining the extent to which they can be regarded as fair and equal. The second part of the chapter discusses Floridi and Cowls' Unified Framework of Five Principles for AI in Society as an important framework for assessing how AI may affect society, including information on how accidental collectives may adhere to or depart from these standards. The complex interactions between AI and fairness are the main topic of the third part of the chapter. In this, I examine the challenges posed by AI algorithms in ensuring fairness and preventing bias within accidental collectives. The focus of this part is on ethical issues related to decision-making procedures, data sources, and algorithmic openness. Finally, the future of accidental collectives over the next 10 to 50 years is explored in the final part of the chapter. Speculating across a spectrum that ranges from extreme AI evolution to more pragmatic and grounded developments, this section explores potential scenarios for the role of accidental collectives in society. The chapter concludes with the suggestion that the more we approach general artificial intelligence, the more accidentality in the generation of collectives lessen.

**Keywords** Ethical AI • Fairness • Social justice • Bias • Future technologies

© The Author(s), under exclusive license to Springer Nature
Switzerland AG 2024
S. Carta, *How Computers Create Social Structures*,
https://doi.org/10.1007/978-3-031-62852-8_5

## Are Accidental Collectives Fair and Equal?

In the previous chapters, we saw how grouping takes place in algorithmic processes. It emerged that much depends on the attributes that users input into the platform to define themselves, and on the way such features are considered by those who design and tweak the sorting and clustering mechanisms in the platform. I suggest we consider the question that titles this section by using a layered approach to individual and group identity. The first level corresponds to the user's own definition of themselves. Each user on a digital platform creates a careful digital version of their identity. The groups generated are authentic because they are chosen directly by the users, but they are not necessarily true, as each user would likely provide an edited version of themselves to fit certain goals within the platform.

The second level is characterised by the assumptions made by those who design and maintain the platform. These are strongly driven by strategic and often commercial decisions within the platform, so we may expect to have invisible forces at play that push the generation of collectives in one direction or another. Examples may include cases where a social media platform may make it easier for users to accidentally come into contact with other users from the same school, gym, or club. Such encounters are perceived as accidental by users, and there may indeed be a degree of randomness in the actual composition of the suggested groups, yet such clustering occurs most likely by design, driven by a clear set of principles underpinning the platform. In other words, the groups created are most likely not a true reflection of that particular reality. They rather offer a very partial and often skewed version of social grouping, which, in a broad sense, is a reflection of the way in which the platform works.

The third level of identity production is intrinsic to the actual sorting and clustering mechanisms. We should look at this layer as a lower level of grouping, where decisions are more related to statistical methods than human intervention. A very illustrative example has been recently provided by Grégoire Martinon in his case of biased automated recruitment (Martinon 2021). In this experiment, the author considers a dataset used to analyse applicants' profiles for a given job. An initial explorative analysis with linear regression indicates that the set contains (or yields) a certain degree of bias, specifically one part of the population is expected to occupy lower job positions with lower pay, compared to the other part. After measuring the degree of fairness, the author implements six different approaches to try to minimise or remove bias entirely. Approaches include using the set as it is ("do nothing"), considering certain user attributes over others ("remove sensitive attribute"), rebalance the weights for the unprivileged groups in the population ("preprocessing by reweighing"), implementing positive discrimination in those cases where the model is uncertain ("post-processing by rejecting"), adding additional penalties in the regression model to mitigate the bias ("in-processing with constraints"), and finally modifying the decision boundary in the clustering space manually to rebalance the outcome ("your strategy"). Martinon illustrates how different approaches to bias reduction result in different outcomes, with arguably none that stands up as the best approach. Each method resolves one part of the problem, emphasising another. Where one type of bias is mitigated (or even eradicated), another surfaces. The author concludes that: "*Algorithmic biases are subtle and ubiquitous. Fairness is hard to achieve and call into question the very notion of "performance". There is no magical metric to optimize nor magical library to solve the problem. And there will never be. Transparency is key Human-in-the-loop is crucial.*" (Martinon 2021).

In light of what we have discussed so far, we may argue that the question in this section of whether accidental collectives are fair and equal is ill-formulated. Accidental collectives are the organic result of an accurate application of statistical methods (at a low level), which informs the design of algorithms that automatically sort and cluster data (at a medium level, with some degree of human intervention), which finally organises users in groups using the characteristics that humans (users) have directly inputted to define themselves within the platform. We could consider equality and fairness at the three levels, with a gradual crescendo of consciousness. At the statistical level, bias is considered a mathematical expression and factored in as a measurable part of the calculations. At the algorithmic level, bias is present and used consciously in some cases, for example, when attributing certain weight to certain attributes that contribute more or less to the combination of groups, as we all as unconsciously. This is when the unconscious bias of the developers may emerge, and this has to do with the educational and cultural background of the developers, the type of data available and used for the training, as well as the way in which the data is trained to yield the desired results. Finally, the highest degree of bias emerges at the user level. When creating their profile on the platform, users tend to generate a very curated version of themselves, omitting certain details and emphasising others. We are all familiar with the idea of digital avatars, a version of ourselves that we design for a specific digital platform to reflect a persona defined by a set of characteristics that we strategically choose. The avatar is partially driven by the ethos of the platform, for example, professionally oriented social media naturally attracts work-related profiles, where academic and professional achievements are emphasised with the aim of networking and wide professional recognition. Other platforms more oriented to social life may drive users to create personas that emphasise social and cultural aspects of their lives, like travelling, cooking, or commenting on restaurant menus. Influencers are probably the highest manifestation of this phenomenon we have today. They develop a business model that revolves around their own persona, with clear opinions on a variety of topics that followers generally find interesting and worth listening to. Looking at these extreme cases, accidental collectives are definitely not fair or equal, for they are the natural result of groups of people organically combined on the basis of a common interest or attraction using a leader-followers model. The influencer portraits a version of reality that is often highly selective and unique, with a very sophisticated lifestyle that, although it strongly resonates with followers, does not reflect the way in which the majority of them conduct their lives. Generally speaking, digital avatars do not offer a fair and equal version of the world, as they are often built with a clear purpose in mind, strategically, and with a careful selection of characteristics with which we want the world to know us.

This upper level is inherently biased, as users generate highly curated versions of themselves through mechanisms that are far from principles of fairness and equality. In fact, we may argue that the principles underpinning the generation of digital avatars work in the opposite direction, for they are driven by the desire for the emergence of the individual over the group using the features of the platform as a stage. Generalising, social media users want to be noticed, followed, commented upon, and ultimately, recognised within their own collectives. To a certain extent, this is a natural reflection of societal and groups dynamics. It is true that different platforms may sway users to emphasise specific characteristics or behave differently depending on the context and the collective they find themselves in, but it is also true that ultimately the users decide upon their own actions with a great degree of freedom within the platform. The point is that if accidental collectives are not entirely equal and fair, this should be considered not within the confined domain of social media but in the wider context of society as one of its many manifestations.

The domain where perhaps more can be done is at the middle level, where algorithmic approaches can be considered in light of a more robust framework for fairness and equality. This is a key level where we have human and machine interaction and where automated routines can be designed in ways in which bias can be more controlled and negative effects can be mitigated. The next section discusses some of these existing frameworks and principles that may be key to improving the level of fairness and equality in the formation of accidental collectives.

## A HELPFUL FRAMEWORK

In their study titled "A Unified Framework of Five Principles for AI in Society" from 2019, Floridi and Cowls suggested five ethical AI core principles (Floridi and Cowls 2019). These guidelines are intended to direct the creation and application of artificial intelligence in a way that is consistent with moral ideals and societal welfare. The work of Floridi and colleagues provide us with very useful guidelines within the context of the generation of accidental collectives.

The first principle is Benefice: Promoting Well-Being, Preserving Dignity, and Sustaining the Planet. This principle hinges on the idea that AI systems should be developed and applied in ways that advance human welfare. Developers and practitioners of AI should strive to produce technologies that benefit both people and society as a whole. This entails making sure that AI programs prioritise human wellbeing, make people's lives better, and advance the general welfare. For instance, AI in healthcare should prioritise patient outcomes and safety, whereas AI in transportation should prioritise reducing accidents and having a minimal impact on the environment. Within the context of our study, AI algorithms should consider social media users' wellbeing, and the sustainable aspects of human activities within the platform.

The Non-Maleficence: Privacy, Security and 'Capability Caution' principle emphasises that AI systems should refrain from harming people or producing unfavourable effects on society. Users and creators of AI systems must be aware of potential dangers and unanticipated outcomes and take precautions to reduce harm. An example could be autonomous vehicles that should be developed to reduce the possibility of accidents and prioritise human safety. AI systems employed in criminal justice should also not reinforce preexisting biases or unfairly target particular populations.

The principle of Autonomy: The Power to Decide (to Decide) addresses the need for AI systems to uphold and support individual autonomy and human agency. Instead of being used to manipulate or control human behaviour, AI should be developed to enable people to make wise decisions. For instance, AI-driven recommendation systems should offer consumers clear and intelligible options so they can choose for themselves what data they want to view or what products they want to buy.

The fourth principle is Justice: Promoting Prosperity, Preserving Solidarity, Avoiding Unfairness, by which AI systems must be created and used in a fair and just manner. This entails taking into account the fair distribution of the advantages and disadvantages of AI technology among various groups and ensuring that no person or community is unjustly disadvantaged. As an illustration, AI algorithms employed in hiring procedures should be developed to prevent discrimination and support equal opportunities for all applicants.

Finally, the Explicability: Enabling the Other Principles through Intelligibility and Accountability principle emphasises the significance of openness and responsibility in AI systems. Intelligent systems must be created in a way that makes it possible to explain their choices and actions. Users and other stakeholders should be able to comprehend and contest the conclusions drawn by AI systems. For instance, AI applied in medical diagnostics should be able to justify its suggestions, assisting patients and healthcare professionals in understanding the diagnosis and available treatments.

## AI AND FAIRNESS

The principles outlined by Floridi and colleagues summarise a large number of studies on algorithmic fairness and AI. This is a very important aspect of Artificial Intelligence and widely studied by many scholars, among which we can include (Dwork et al. 2012; Kim et al. 2017; Mittelstadt et al. 2016; Selbst and Barocas 2018; Wachter et al. 2017; Žliobaité 2017). It is worth noting the important work has been carried out on national and international level with overall standards and guidelines. For example, the UK recently produced a comprehensive policy paper on the use of fair and equitable AI (Secretary of State for Science, Innovation and Technology 2023) as a part of the UK national AI Strategy. Similarly other countries like France released in the past few years national policy documents on fair use of AI as a part of their strategic development, for example with Artificial Intelligence, Big Data and Fundamental Rights (Country Research France 2020). The European Union recently published a detailed a Regulatory framework proposal on artificial intelligence (European Commission 2023; Niklas and Dencik 2020), as well as the USA with the National Institute of Standards and Technology (NIST)'s Towards a Standard for Identifying and Managing Bias in Artificial Intelligence (Schwartz et al. 2022), India with Standard for Fairness Assessment and Rating of Artificial Intelligence Systems (Telecommunication Engineering Centre 2023; Agarwal and Agarwal 2023; Agarwal et al. 2023). Large and global organisations like IEEE are actively working on standards framework, including the IEEE global mission, "to ensure every stakeholder involved in the design and development of autonomous and intelligent systems is educated, trained, and empowered to prioritize ethical considerations so that these technologies are advanced for the benefit of humanity." (IEEE 2023; Chatila and Havens 2019). Others include the work conducted by group behind the ACM Conference on Fairness, Accountability, and Transparency (*ACM FAccT* 2023) whose mission is to combine technical solutions with important problems regarding the distribution of power, perverse effects, and redistribution of welfare; and to provide a foundation for study on fairness, accountability, and transparency in current legal requirements (*ACM FAccT* 2023).

Generally speaking, there are a number of aspects that emerge from this collective work that is taking place globally. The notion of responsibility appears as one of the main points of discussion. Those who design, develop and implement algorithms should be held responsible for their choices and any unintended consequences they may have on individuals and society. This entails actively striving to mitigate negative impacts and accounting for the algorithmic decision-making's broader societal implications. Responsible algorithm design requires extensive testing, validation, and continuing monitoring to identify and address biases, errors, and other potential issues that may appear throughout algorithmic decision-making processes.

Another key point is around Explainability. This highlights how important it is to make algorithms transparent and comprehensible. Users and other interested parties should be able to understand the factors influencing algorithmic outputs and decision-making processes when AI systems are developed. Users are more inclined to trust the system and are better equipped to identify biases or mistakes if they are aware of how assessments are made. Transparency is crucial, especially in high-risk industries like healthcare and criminal justice.

The idea of Accuracy addresses the need of making sure algorithms are reliable and produce accurate results. This necessitates stringent algorithm development, testing, and validation methods in order to decrease mistakes and improve system performance. High accuracy is crucial, particularly in delicate applications where unreliable results could have a significant influence on people's lives or corporate decisions.

AI systems need to be auditable (Auditability). Procedures must be put in place to allow outside parties to review and assess algorithms. Thanks to auditable algorithms, external academics, authorities, and the general public can evaluate the system's fairness, accountability, and overall success. Accountability checks are made possible by this process, which also aids in the identification of any biases or detrimental impacts and encourages algorithm developers to be more receptive to any issues.

Finally, we have Fairness. This idea underlines the importance of considering how algorithms could affect different demographic groups. Fairness demands that results that are unfair or discriminatory due to protected characteristics like race, gender, age, or ethnicity be avoided. Fair algorithms work to lessen differential effects, guarantee fair treatment of different subpopulations, and stop societal injustices from continuing.

Overall, any intelligent system using AI should have (i) a clear **purpose and scope** (with the algorithm's intended use as well as its range of potential applications, outlining the duties it is created to carry out and the data it handles), (ii) document its **data sources and collection method** (describing the sources utilised to develop and run the algorithm, as well as any biases or constraints that may have been in the data), (iii) **highlight potential impacts** (assessing the potential positive and negative effects that the implementation of the algorithm may have on specific people, groups of people, and society at large), (iv) include **strategies for mitigating any negative effects** (including constant monitoring, algorithmic openness, and fairness concerns), (v) **actively involve relevant stakeholder** (recognising the contributions of pertinent stakeholders, considering their viewpoints and issues when developing and deploying the algorithm).

## ACCIDENTAL COLLECTIVES IN THE FUTURE

In this section I explore how the future of the automatic and accidental generation of collectives may look like in the next 10–50 years. I speculate within a range that goes from an extreme evolution of artificial intelligence to a more realistic and humbler version of possible developments.

In order to build an extreme scenario, we will use two ideas: the Master Algorithm (Domingos 2015) and the solution to the P Vs NP problem (Fortnow 2013). Both ideas concern with the convergence of technology towards a general artificial intelligence, where computers would be able to match and possibly surpass human intelligence. These ideas are somewhat in line with other notions, such as Superintelligence, where artificial intelligence would become more capable and powerful than that of humans (Bostrom 2017) and Singularity (Kurzweil 2005), where human and machine intelligences would merge, defining a radical next step in the human evolution.

Pedro Domingos' idea of the Master Algorithm suggests the invention of an algorithm that can successfully learn from any type of data and resolve any sort of issue. This concept lies on the idea that there is a single, all-inclusive algorithm that can unify the many machine learning and AI frameworks. The development of such a universal learning algorithm will advance AI greatly and have a big impact on many different fields and industries. The Master Algorithm's foundation is made up of several approaches that, combined would allow this algorithm to exist. Domingos calls them the five tribes of Machine Learning (Domingos 2015:51) and they include Symbolists: Symbolic AI methods focus on representing knowledge by utilising logical concepts and symbolic representations. These algorithms are designed to assess knowledge bases and arrive at logical conclusions. Connectionist AI techniques, often known as neural networks, are based on how the human brain is organised and focus on learning from data by altering the strength of connections between artificial neurons. Evolutionaries: The main sources of inspiration for evolutionary AI techniques are genetics and natural evolution. These algorithms use evolutionary theory and genetic programming to identify the most effective solutions to complex problems. Bayesian AI approaches use Bayesian probability theory to reason under uncertainty and generate probabilistic predictions based on the available data. Analyzers: Analogical AI methods draw conclusions about the future by drawing comparisons between the past and the present.

The second idea that can help us identifying useful scenarios for the future of collectives comes from Lance Fortnow's book The Golden Ticket: P, NP, and the Search for the Impossible (Fortnow 2013).

The question of whether P Vs NP is one of the most difficult and significant mathematical problems of the millennium and is listed as one of the six unsolved mathematical problems of the Clay Institute (Clay Mathematics Institute 2023). Within the context of this problem, P stands for polynomial time, while NP stands for nondeterministic polynomial time. The two terms describe how long it takes an algorithm to solve a specific task. There are issues that, when solved in a P time (or, more precisely, when they fall into a P complexity class), an algorithm can solve in an effective and reasonably quick manner. A problem is classified as NP if an algorithm can solve it, but only after an exponential time (a very long time opposed to a polynomial time). However, it is simple to verify the answer to such NP issues. This indicates that it is generally simple to confirm the solution's accuracy if one can see it, let's say intuitively. However, determining whether a problem that is quickly solvable (i.e., solvable in polynomial time) is also quickly verifiable is not a trivial question.

Fortnow's book describes a scenario where P is actually equal to NP, that is to say we can resolve any possible problems we may have now and in the future relatively easily. This includes cure any disease, make extraordinary advancement in science and society. Through the fictional Urbana Algorithm, Fortnow presents a beautiful world (Fortnow 2013:11) where we computers can help us eradicating cancer, find any criminal in a matter of minutes, automate tasks for humanities, create global prosperity and give everyone a better version of the world.

In both scenarios suggested by Domingos and Fortnow social collectives would be generated through a higher computing intelligence. This would mean that, hypothetically, there will be no trace of unfairness and bias in the data, all groups would be created with utter trustworthiness and equality. People would entirely delegate the solution of these problems to algorithms that would have surpassed human intelligence. As such, humans would not be able to understand the mechanisms of such systems, so the only option left would be to entirely trust these superior systems. This scenario would have an extreme version of what we now call black box approach, where we know the input of an intelligent system and we can assess the output, but we have little comprehension of how the system reach the conclusions. In the utopic scenario described here, we would not probably be even in control of the inputs, as such advanced systems would autonomously select them to guarantee the desired results (a society with no crimes and no diseases).

These scenarios would mean the end of the accidentality in the generation of collective. Following the assumption that a machine of superior intelligence would have utter control over the world, yielding infallible results, it would be safe to assume that in such a perfectly controlled reality accidents should not occur. There would be no glitches, nor will there be unintended outcomes in any computation. The formation of social groups would be entirely accurate and would leave no room for emergent manifestations of human behaviour. All the rich unpredictability and "shifts, uncertainties, and mess which are real life" (Sennett 2012) would have been flattened in exchange for a perfect outcome.

The second and more realistic possible evolution of social collectives would be the result of the gradual development of intelligent systems that we are currently witnessing today. The development of AI and automated systems in general has been increasing for the past decades. There have been changes in pace of such developments, but there is a general trend that indicates growth both in quantity (of tools, approaches and methods) and capability, with AI systems able to get closer to general artificial intelligence. The growing development in AI studies is evidenced in the increasing number of publications of studies in high-impact academic journals. We can see a significant spike in publications in the past 5 years (Tang et al. 2020). There is also evidence (Roser 2022) to suggest that AI models are reaching a point where they are performing better at tests compared to human results in a number of fields, including reading and comprehension, language understanding, and in speech, images and handwriting recognition (Roser 2022). AI models are then being developed with steady growth. One the most significant recent development in this field is probably the introduction of diffusion models and large language models (LLMs). These tools underpin the recent explosion of general-purpose AI platforms like ChatGPT and the like, where a wider section of society started using AI models as an aid to repetitive and laborious tasks, including summarising texts, writing detailed paragraphs from bullet point lists, paraphrasing, describing concepts (Malik et al. 2023), or even establishing human-computer conversations (Gottlieb et al. 2023).

Such advancement would suggest that we will be witnessing a gradual and steady improvement in many fields and applications of intelligent systems and AI, including in social network analysis. This would mean that the limits that we discussed in the previous chapters that characterise the formation of groups in social media platforms will gradually diminish. We may argue that most of the problems that may emerge from accidental formation of social groups can be related to the dedicated single-objective functions that drive the clustering and grouping algorithms. The accidentality is a consequence of a single task which may generate collectives as a byproduct. For example, users in a platform may end up clustered together as the un-scripted objective of maximising features like affinity among users, common interests or geographical location. The more the artificial intelligence driving the clustering of users in a platform is narrow (i.e., good at solving one particular problem), the higher the chances are for a group to be created accidentally. Conversely, the more general the AI is, the less the accidentality of the collective is. Under this assumption, it would be safe to say that, as AI moves gradually from narrow to general, collectives will be increasingly less accidental.

## REFERENCES

ACM FAccT. 2023. Available at: https://facctconference.org/ [Accessed: 9 August 2023].

Agarwal, A. and Agarwal, H. 2023. A seven-layer model with checklists for standardising fairness assessment throughout the AI lifecycle. *AI and Ethics*, pp. 1–16.

Agarwal, A., Agarwal, H. and Agarwal, N. 2023. Fairness Score and process standardization: framework for fairness certification in artificial intelligence systems. *AI and Ethics* 3(1), pp. 267–279.

Bostrom, N. 2017. *Superintelligence*. Dunod.

Chatila, R. and Havens, J.C. 2019. The IEEE global initiative on ethics of autonomous and intelligent systems. *Robotics and well-being*, pp. 11–16.

Clay Mathematics Institute. 2023. *P vs NP*. Available at: https://www.claymath. org/millennium/p-vs-np/ [Accessed: 10 August 2023].

Country Research France. 2020. *Artificial Intelligence, Big Data and Fundamental Right*. European Union Agency for Fundamental Rights (FRA). Available at: https://fra.europa.eu/sites/default/files/fra_uploads/fra-ai-project-france-country-research_en.pdf.

Domingos, P. 2015. *The master algorithm: How the quest for the ultimate learning machine will remake our world*. Basic Books.

Dwork, C., Hardt, M., Pitassi, T., Reingold, O. and Zemel, R. 2012. Fairness through awareness. In: *Proceedings of the 3rd innovations in theoretical computer science conference*. pp. 214–226.

European Commission. 2023. *Regulatory framework proposal on artificial intelligence | Shaping Europe's digital future.* Available at: https://digital-strategy.ec.europa.eu/en/policies/regulatory-framework-ai [Accessed: 9 August 2023].

Floridi, L. and Cowls, J. 2019. A Unified Framework of Five Principles for AI in Society. *Harvard Data Science Review* 1(1). Available at: https://hdsr.mitpress.mit.edu/pub/l0jsh9d1/release/8 [Accessed: 8 August 2023].

Fortnow, L. 2013. *The golden ticket: P, NP, and the search for the impossible.* Princeton University Press.

Gottlieb, M., Kline, J.A., Schneider, A.J. and Coates, W.C. 2023. ChatGPT and conversational artificial intelligence: Friend, foe, or future of research? *The American Journal of Emergency Medicine* 70, pp. 81–83.

IEEE. 2023. *The IEEE Global Initiative on Ethics of Autonomous and Intelligent Systems.* Available at: https://standards.ieee.org/industry-connections/ec/autonomous-systems/ [Accessed: 9 August 2023].

Kim, B., Malioutov, D.M., Varshney, K.R. and Weller, A. 2017. Proceedings of the 2017 ICML workshop on human interpretability in machine learning (WHI 2017). *ArXiv e-prints*, p. arXiv-1708.

Kurzweil, R. 2005. The singularity is near. In: *Ethics and emerging technologies.* Springer, pp. 393–406.

Malik, T. et al. 2023. "So what if ChatGPT wrote it?" Multidisciplinary perspectives on opportunities, challenges and implications of generative conversational AI for research, practice and policy. *International Journal of Information Management* 71, p. 102642.

Martinon, G. 2021. *Tutorial : breaking myths about AI fairness. The case of biased automated recruitment.* Available at: https://towardsdatascience.com/tutorial-breaking-myths-about-ai-fairness-the-case-of-biased-automated-recruitment-9ee9b2ecc3a [Accessed: 8 August 2023].

Mittelstadt, B.D., Allo, P., Taddeo, M., Wachter, S. and Floridi, L. 2016. The ethics of algorithms: Mapping the debate. *Big Data & Society* 3(2), p. 2053951716679679.

Niklas, J. and Dencik, L. 2020. European artificial intelligence policy: Mapping the institutional landscape.

Roser, M. 2022. *The brief history of artificial intelligence: The world has changed fast – what might be next?* Available at: https://ourworldindata.org/brief-history-of-ai [Accessed: 10 August 2023].

Schwartz, R., Vassilev, A., Greene, K., Perine, L., Burt, A. and Hall, P. 2022. Towards a standard for identifying and managing bias in artificial intelligence. *NIST special publication* 1270(10.6028).

Secretary of State for Science, Innovation and Technology. 2023. *A pro-innovation approach to AI regulation.* Available at: https://www.gov.uk/government/publications/ai-regulation-a-pro-innovation-approach/white-paper [Accessed: 9 August 2023].

Selbst, A.D. and Barocas, S. 2018. The intuitive appeal of explainable machines. *Fordham L. Rev.* 87, p. 1085.

Sennett, R. 2012. No one likes a city that's too smart. *The Guardian.* 4 December. Available at: https://www.theguardian.com/commentisfree/2012/dec/04/smart-city-rio-songdo-masdar [Accessed: 10 August 2023].

Tang, X., Li, X., Ding, Y., Song, M. and Bu, Y. 2020. The pace of artificial intelligence innovations: Speed, talent, and trial-and-error. *Journal of Informetrics* 14(4), p. 101094.

Telecommunication Engineering Centre. 2023. *Fairness Assessment and Rating of Artificial Intelligence Systems.* Available at: https://tec.gov.in/ai-fairness [Accessed: 9 August 2023].

Wachter, S., Mittelstadt, B. and Floridi, L. 2017. Transparent, explainable, and accountable AI for robotics. *Science robotics* 2(6), p. eaan6080.

Žliobaitė, I. 2017. Measuring discrimination in algorithmic decision making. *Data Mining and Knowledge Discovery* 31(4), pp. 1060–1089.

# INDEX

© The Author(s), under exclusive license to Springer Nature
Switzerland AG 2024
S. Carta, *How Computers Create Social Structures*,
https://doi.org/10.1007/978-3-031-62852-8